A General Theory of Love

THOMAS LEWIS, M.D., is an assistant clinical professor of psychiatry at the University of California, San Francisco, School of Medicine, and a former associate director of the Affective Disorders Program there. Dr. Lewis currently divides his time between writing, private practice, and teaching at the UCSF medical school. He lives in Sausalito, California.

FARI AMINI, M.D., is a professor of psychiatry at the UCSF School of Medicine. Born and raised in Iran, he graduated from medical school at UCSF and has served on the faculty there for thirty-three years. He has also been on the faculty at the San Francisco Psychoanalytic Institute since 1971, and served as its president in 1981. Dr. Amini is married, has six children, and lives in Ross, California.

RICHARD LANNON, M.D., is an associate clinical professor of psychiatry at the UCSF School of Medicine. In 1980, Dr. Lannon founded the Affective Disorders Program at UCSF, a pioneering effort to integrate psychological concepts with the emerging biology of the brain. Dr. Lannon is married and the father of two; he lives in Greenbrae, California.

Drs. Lewis, Amini, and Lannon have been working together since 1991. Each comes from a different generation of psychiatrists: Dr. Amini from one in which psychoanalysis reigned unchallenged, Dr. Lannon from the era that first saw the use of psychoactive medication in treating emotional illness, and Dr. Lewis from the recent generation of psychiatrists who trained during the collision of psychodynamics with neuroscience. Dissatisfied with the standard accounts of the mind, they combined their energies to construct alternative paradigms. Their collaboration has generated academic papers and numerous presentations for psychiatric professionals. Perhaps most important, their partnership has spawned the most precious outcomes of collaboration: creativity, pleasure, and friendship.

A
General Theory
of Love

Thomas Lewis, M.D.,

Fari Amini, M.D.,

Richard Lannon, M.D.

VINTAGE BOOKS

A Division of Random House, Inc.

New York

FIRST VINTAGE EDITION, JANUARY 2001

Copyright © 2000 by T. Lewis, F. Amini, and R. Lannon

All rights reserved under International and Pan-American Copyright
Conventions. Published in the United States by Vintage Books, a division of
Random House, Inc., New York, and simultaneously in Canada by Random
House of Canada Limited, Toronto. Originally published in hardcover in the
United States by Random House, Inc., New York, in 2000.

Vintage and colophon are registered trademarks
of Random House, Inc.

Owing to limitations of space, acknowledgments of permission
to quote from previously published materials will be
found following the index.

The Library of Congress has cataloged the
Random House edition as follows:
Lewis, Thomas.
A general theory of love / Thomas Lewis, Fari Amini, Richard Lannon.
p.   cm.
ISBN: 0-375-50389-7
Includes bibliographical references (p. [241]–254) and index.
I. Love.   2. Love—Physiological aspects.   I. Title.
BF575.L8 L49 2000
152.4'I—dc2I 99-049930
CIP

Vintage ISBN: 0-375-70922-3

Book design by Mercedes Everett and Lisa Motzkin

www.vintagebooks.com

Printed in the United States of America
10   9   8

Once you have flown,

you will walk the Earth

with your eyes turned skyward;

for there you have been,

there you long to return.

—Leonardo da Vinci

PREFACE

What is love, and why are some people unable to find it? What is loneliness, and why does it hurt? What are relationships, and how and why do they work the way they do?

Answering these questions, laying bare the heart's deepest secrets, is this book's aim. Since the dawn of our species, human beings in every time and place have contended with an unruly emotional core that behaves in unpredicted and confusing ways. Science has been unable to help them. The Western world's first physician, Hippocrates, proposed in 450 B.C. that emotions emanate from the brain. He was right—but for the next twenty-five hundred years, medicine could offer nothing further about the details of emotional life. Matters of the heart were matters only for the arts— literature, song, poetry, painting, sculpture, dance. Until now.

The past decade has seen an explosion of scientific discoveries about the brain, the leading edge of a revolution that promises to change the way we think about ourselves, our relationships, our children, and our society. Science can at last turn its penetrating gaze on humanity's oldest questions. Its revelations stand poised to shatter more than a few modern assumptions about the inner workings of love.

Traditional versions of the mind hold that Passion is a troublesome remnant from humanity's savage past, and the intellectual subjugation of emotion is civilization's triumph. Logical but dubious derivations follow: emotional maturity is synonymous with emotional restraint. Schools can teach children missing emotional skills just as they impart the facts of geometry or history. To feel better, outthink your stubborn and recalcitrant heart. So says convention.

In this book, we demonstrate that where intellect and emotion clash, the heart often has the greater wisdom. In a pleasing turnabout, science—Reason's right hand—is proving this so. The brain's ancient emotional architecture is not a bothersome animal encumbrance. Instead, it is nothing less than the key to our lives. We live immersed in unseen forces and silent messages that shape our destinies. As individuals and as a culture, our chance for happiness depends on our ability to decipher a hidden world that revolves—invisibly, improbably, inexorably—around love.

From birth to death, love is not just the focus of human experience but also the life force of the mind, determining our moods, stabilizing our bodily rhythms, and changing the structure of our brains. The body's physiology ensures that relationships determine and fix our identities. Love makes us who we are, and who we can become. In these pages, we explain how and why this is so.

During the long centuries when science slumbered, humanity relied on the arts to chronicle the heart's mysterious ways. That accumulated wisdom is not to be disdained. This book, while traveling deep into the realm of science, keeps close at hand the humanism that renders such a journey meaningful. The thoughts of researchers and empiricists join those of poets, philosophers, and kings. Their respective starting points may be disparate in space, time, and temperament, but the voices in this volume rise and converge toward a common goal.

Every book, if it is anything at all, is an argument: an articulate arrow of words, fledged and notched and newly anointed with sharpened stone, speeding through paragraphs to its shimmering target. This book—as it elucidates the shaping power of parental devotion, the biological reality of romance, the healing force of communal connection—argues for love. Turn the page, and the arrow is loosed. The heart it seeks is your own.

CONTENTS

A General Theory of Love

One

THE HEART'S CASTLE

Science joins the search for love

*Two girls discover
the secret of life
in a sudden line of
poetry.*

*I who don't know the
secret wrote
the line. They
told me*

*(through a third person)
they had found it
but not what it was
not even*

*what line it was. No doubt
by now, more than a week
later, they have forgotten
the secret,*

*the line, the name of
the poem. I love them
for finding what
I can't find,*

*and for loving me
for the line I wrote,*

and for forgetting it
so that

a thousand times, till death
finds them, they may
discover it again, in other
lines

in other
happenings. And for
wanting to know it,
for

assuming there is
such a secret, yes,
for that,
most of all.

—Denise Levertov, "The Secret"

Some might think it strange that a book on the psychobiology of love opens with a poem, but the adventure itself demands it. Poetry transpires at the juncture between feeling and understanding—and so does the bulk of emotional life. More than three hundred years ago, the French mathematician Blaise Pascal wrote, *The heart has its reasons whereof Reason knows nothing.* Pascal was correct, although he could not have known why. Centuries later, we know that the neural systems responsible for emotion and intellect are separate, creating the chasm between them in human minds and lives. The same rift makes the mysteries of love difficult for people to penetrate, despite an earnest desire to do so. Because of the brain's design, emotional life defeats Reason much as a poem does. Both retreat from the approach of explication like a mirage on a summer's day.

Although the nature of love is not easy to define, it has an intrinsic order, an architecture that can be detected, excavated, and explored. Emotional experience, in all its resplendent complexity, cannot emerge *ex vacuo:* it must originate in dynamic neural systems humming with physiologic machinations as specific and patterned as they are intricate. Because it is part of the physical universe, love has to be *lawful.* Like the rest of the world, it is governed and described by principles we can discover but cannot change. If we only knew where and how to look, we should be able to find emotional laws whose actions a person could no more resist than he could the force of gravity if he fell off a cliff.

Locating love's precepts is a daunting task. Every conception of love inevitably depends on a view of the broader totality of the emotional mind. Until the most recent snippet of human history, however, a science of the mind did not exist. Classical Greek disciplines included geometry, astronomy, medicine, botany—but no conception of human emotions that could claim more credibility than their contemporaneous and vivid myths. That empirical emptiness endured for thousands of years. Philosophers expounded and debated on emotional life—four bodily humors here, demonic possession there—but the world waited until the end of the nineteenth century A.D. for systematic investigation into feelings and passion.

When scientific attention first turned to the heart's mysteries, the technologies essential to solving them were inconceivable. At the end of the nineteenth century, a handful of thinkers—Sigmund Freud, William James, Wilhelm Wundt—worked on assembling the earliest scientific accounts of human mental faculties. Brilliant pioneers though they were, they could know nothing about the mind's physicality, about the minuscule neural mechanisms that combine and conspire to create the stuff of mental life—sights, sounds, thoughts, ambitions, feelings. Love's secrets remained buried within the most impenetrable treasure chest the world has ever known: a tangle of a hundred billion cells, whose innumerable electrical currents and chemical signals come together to create a single, living human brain.

From the beginning of the twentieth century to its end, influential accounts of love included no biology. It has been said that neurotics build castles in the sky, while psychotics live in them, and psychiatrists collect the rent. But it is the psychiatrists and psychologists who have been living within a palace of theory suspended over a void. When they built their understanding of the emotional mind, the brain was a cipher. The foundations of their edifice had to be fashioned out of the only substance in plentiful supply—the purest speculation.

The first explorers of humanity's passions met that challenge with bold invention. In a sanctuary safe from refutation, they conjured up mental contraptions and metaphors that had no physical referent. Sigmund Freud was not the only dreamer who sketched an impressionistic vision of the mind, but he was the most relentless in crediting his concoction with a solidity it could not possibly possess. And so the towers and walls of the Freudian citadel sprang into midair, where they remain: the looming turret of the censoring superego, the lofty arches of insight, the squat dungeon of the id. Despite the insubstantial base, that old model of emotional life cast a long shadow. Freud is delivered anew to each generation. His conclusions permeate our culture in a multitude of ways, and his assumptions have endured for so many years that they are mistaken for fact.

The cultural atmosphere in Freud's time was suffused with suspicion about the moral and physical hazards of masturbation. Freud, who disapproved of masturbation for the duration of his

life, was convinced that onanism and coitus interruptus were responsible for anxiety, lassitude, a plethora of hysterical symptoms—the emotional dysfunctions of his day. Next he concluded
that childhood sexual seduction was the real culprit; then his focus
shifted to youthful *fantasies* of copulation with parents. When his
clinical encounters revealed that most patients denied all varieties
of precocious eroticism, Freud did not question his original conviction. He concluded that patients did not remember young, sen-
suous adventures because the mind had spirited memories out of
consciousness. When he sifted through his patients' symptoms and
dreams, he believed he could see cleverly encrypted clues pointing
to a dark sexual history—the same one, he failed to notice, that he
had envisioned from the outset.

This prototype of the emotional mind contains familiar
Freudian machinery: desire's cauldron bubbling beneath the surface
of awareness; the sunlit quotidian existence of the self, incognizant
of lurking nether regions; and the healing power of insight into a
sinister erotic past that, by definition, *has* to be there. This account
of humanity's heart binds love inextricably to sexual pleasure and
perversion—indeed, it holds that love is but a convoluted representation of forbidden, repellent, incestuous urges. For his emblem
and standard-bearer, Freud scanned the roster of the Greek theater
and chose Oedipus—who, cursed by the gods, an inadvertent per-
vert and parricide, blinds himself and wanders in misery. The
adopted story's transfigured moral is that the civilizing forces of
reason and intellect must reign if humanity's bestial nature is not
to descend toward unspeakable horror.

"Man is a credulous animal and must believe *something*," wrote
Bertrand Russell. "In the absence of good grounds for belief, he
will be satisfied with bad ones." Wherever and whenever they are,
people vastly prefer any explanation (however flawed or implausible) to none. When Freud announced that he had plumbed once

and for all the inky depths of human passions, a world desirous of reassuring certainty flocked to his vision.

As in a dictatorship, however, embracing the end of anarchy came at a price. Freud's logic was a veritable Möbius strip of circularity. When patients complied with his insistence that they remember early sexual material, he called them astute; when they did not, he said they were resisting and repressing the truth. (Equating denial with confession is a versatile, albeit ignoble, tool that has served diverse enterprises, from the burning of Salem's witches to the persecutions of the Inquisition.) Today, Freud's conclusions are said to be regularly validated by the practice of insight-oriented psychotherapy—but that activity is the sole purview of those who have already accepted the tenets they subsequently purport to confirm. Such revolving door reasoning could corroborate any proposition, no matter how faulty.

Psychoanalytic concepts captivated popular culture as have no other ideas about humanity's mind and heart. But the Freudian model belongs to a prescientific era in the search to unravel the enigmas of love. The demise of such mythologies is always probable. As long as the brain remained a mystery, as long as the physical nature of the mind remained remote and inaccessible, an evidential void permitted a free flow of irrefutable statements about emotional life. As in politics, the factor determining the longevity and popularity of these notions was not their veracity but the energy and wit devoted to promoting them.

In the years when unrestrained presumptions about the mind roamed free, outlandish claims piled up like election year promises. Seizures are covert expressions of orgasmic ecstasy, one theory maintained. Children who lag in their reading and writing skills are exacting revenge on parents who expelled them from the marital bed. A migraine headache discloses sexual fantasies of defloration. All of these colorful assertions were living on time borrowed from the prevailing scientific ignorance about the brain.

Because we three are clinicians, we must answer to the daily demands of pragmatism. The purpose behind discerning the nature of love is not to satisfy ivory tower discussions or to produce fodder for academic delectation. Instead, as our work makes all too clear, the world is full of live men and women who encounter difficulty in loving or being loved, and whose happiness depends critically upon resolving that situation with the utmost expediency. However inelegant or mythological a model of the mind might be, if we found it clinically effective—if we could use it to help people know their own hearts—we would be loath to reject it.

When we sought to make use of the Freudian model and its numerous offshoots, however, we discovered that efficacy was not among the model's advantages. When each of us came to grapple with the emotional problems of our patients, we saw that the old models provide diagrams to a territory that cannot be found anywhere within a real person. Our patients never behaved as predicted. They did not benefit from what the models prescribed, and what did help them had never been taught to us. Unless we stretched and contorted it past the breaking point, that framework for understanding emotional life failed to elucidate the stories of the patients we met in our offices every day. And so we sought elsewhere for clues to the heart's perplexing conundrums.

The science of the emotional mind got off to a slow start in the first half of the twentieth century, but in the latter half it found a second and adventitious wind. While French doctors searched for antihistamines, they created antipsychotic medications. Drugs for tuberculosis were observed to improve mood, and a few short chemical steps later, antidepressants blossomed. An Australian accidentally discovered that lithium makes guinea pigs docile, and in so doing he stumbled upon a treatment for manic depression. Tiny molecules, when ingested and transported to the brain, were capable of erasing delusions, removing depression, smoothing out mood swings, banishing anxiety—how could one square *that* with the supposed preeminence of repressed sexual urges as the cause of all matters emotional?

In the 1990s, the collision of pharmacological efficacy with psychoanalytic explanations all but reduced the latter to flinders. At the same time, the displacement of this dominant paradigm left all of us without a coherent account of our lives and loves. Freud's collapse in the last decade of the twentieth century has rendered our yearnings, desires, and dreams, if not inexplicable, then at least unexplained.

Although science has risen to take its place as Freud's successor, it has not been able to sketch a framework for love that is both sound and habitable. Two persistent obstacles block the way.

First, a curious correlation has prevailed between scientific rigor and coldness: the more factually grounded a model of the mind, the more alienating. Behaviorism was the first example: brandishing empiricism at every turn, it was thoroughly discomfiting in its refusal to acknowledge such staples of human life as thought or desire. Cognitive psychology bristled with boxes and arrows linking perception to action and had nothing to say about the unthinking center of self that people most cherish. Evolutionary psychology has shed welcome light on the mind's Darwinian debts, but the model declaims as illusions those features of human life lacking an obvious survival advantage—including friendship, kindness, religion, art, music, and poetry.

Modern neuroscience has been equally culpable of propagating an unappealing and soulless reductionism. If the psychoanalysts spun an intangible castle in the air for humanity to inhabit, neuroscience has delivered a concrete hovel. Is every mood or manner best understood as the outcome of molecular billiard balls caroming around the cranium? When emotional problems arise, is a steady diet of Ritalin for children and Prozac for adults to be our only national response? If a woman loses her husband and becomes depressed, does her sorrow *signify*, or is she just a case of chemistry gone awry? Science is a newcomer to the business of defining human nature, but thus far it has remained inimical to humanism. Seekers of meaning are turned away at the door.

The second impediment to a wholly scientific description of love is the dearth of hard data. Systematic investigation holds out enticing promises for those who wish to understand the brain—and what empiricism gives with one generous gesture, it takes back with another. Despite galactic strides in technology, brain science remains a frustrating collection of pillow-soft hints, bulging with ambiguity. These intimations may point in the right direction, but they will not take us with clean finality to conclusiveness. Science has come far on the path to understanding the brain, but that road stretches on to the horizon. The student of love still confronts a venerable relationship between certainty and utility in matters of the heart: only a few things worth knowing about love can be proven, and just a few things amenable to proof are worth knowing at all.

When he ventures into love's domain, the uncompromising empiricist is left with little to discuss. A child's fierce and inarticulate longing for his parents, the torrential passion between young lovers, any mother's unshakable devotion—all are elusive vapors that mock objectivity's earnest attempt to assign them to *this* gene or *that* collection of cells. Someday, perhaps, everything will be known, but that day beckons from an unimaginable distance. And yet without *some* tethers to verifiable facts, anybody can spin limitless high-blown fancies about love that have the same evidential status as the emanations of a Ouija board.

If empiricism is barren and incomplete, while impressionistic guesswork leads anywhere and everywhere, what hope can there be for arriving at a workable understanding of the human heart? In the words of Vladimir Nabokov, there can be no science without fancy and no art without facts. Love emanates from the brain; the brain is physical, and thus as fit a subject for scientific discourse as cucumbers or chemistry. But love unavoidably partakes of the personal and the subjective, and so we cannot place it in the killing jar and pin its wings to cardboard as a lepidopterist might a prismatic butterfly. In spite of what science teaches, only a delicate admixture of evidence and intuition can yield the truest view of the emotional mind. To slip between the twin dangers of empty reductionism and baseless credulity, one must balance a respect for proof with a fondness for the unproven and the unprovable. Common sense must combine in equal measure imaginative flight and an aversion to orthodoxy.

While science provides a remarkably serviceable tool for exploring and defining the natural world, human beings come equipped with an older means of discerning the nature of the hearts around them. That second way is every bit as influential as logic—in many circumstances, considerably more so. This book imparts the legitimacy and necessity of both methods of reading emotional secrets—a friend's, a partner's, a child's, your own.

For years, the three of us combed the neuroscientific literature looking for the lustrous facts that could illuminate relatedness, for the studies that could unravel the knots and untwine the fibers of the ties that bind. We searched, in short, for the science of love. Finding no such system in our own field, we went hunting in other disciplines. Before we were through scavenging, we had gathered together elements from neurodevelopment, evolutionary theory, psychopharmacology, neonatology, experimental psychology, and computer science.

Although this book traffics in those scientific discoveries, we cannot endorse the myopic assumption that academic papers hold the key to the mysteries of love. Human lives form the richest repository of that information. Those who attempt to study the body without books sail an uncharted sea, William Osler observed, while those who only study books do not go to sea at all. And so, wherever possible, we compared what research had to say against the emotional experience of our patients, our families, and ourselves.

After several years of cross-pollination from a panoply of disciplines, the interdisciplinary maelstrom coalesced. We began to think of love and to describe it to one another in terms we had never heard. A revolutionary paradigm assembled itself around us, and we have remained within it ever since. Within that structure, we found new answers to the questions most worth asking about human lives: what are feelings, and why do we have them? What are relationships, and why do they exist? What causes emotional pain, and how can it be mended—with medications, with psychotherapy, with both? What is therapy, and how does it heal? How should we configure our society to further emotional health? How should we raise our children, and what should we teach them?

The investigation of these queries is not just an intellectual excursion: people *must* have the answers to make sense of their lives. We see the need for this knowledge every day, and we see the bitter consequences of its lack. People who do not intuit or respect the laws of acceleration and momentum break bones; those who do not grasp the principles of love waste their lives and break their hearts. The evidence of that pain surrounds us, in the form of failed marriages, hurtful relationships, neglected children, unfulfilled ambitions, and thwarted dreams. And in numbers, these injuries combine to damage our society, where emotional suffering and its ramifications are commonplace. The roots of that suffering are often unseen and passed over, while proposed remedies cannot succeed, because they contradict emotional laws that our culture does not yet recognize.

Those laws are written in stone somewhere within the heart, regardless of how long they manage to elude discovery. And given the microscopic maze wherein such secrets dwell, centuries may pass before the brain yields up its last mysteries. None of us will live to see beyond the dawn of that revelatory age.

In these pages, we take up the challenge that science puts within reach today—exploring the nature of love, drawing upon imagination, invention, and the ascendant scientific knowledge that biotechnology places at our disposal. By design, we have not produced a comprehensive encyclopedia of brain science. No multi-lettered neuroanatomical diagrams lurk within these pages. We have set out not to map the mind in numbing detail, but to lead an agile reconnaissance over landscapes that lie hidden within the human soul.

As we do so, we will travel afield from what many people consider the proper territory of the psyche. Before we are through, we will touch upon the mewling of lost puppies, the mathematics of memory, the marital fidelity of prairie dogs, and the facial expressions in the South Pacific. We will consider the child-rearing experiment of a medieval emperor, psychotherapeutic techniques, the intuitive genius of newborns, and why people hold hands at the movies. We will ask why families exist, what feelings are, and what love is not; how blind babies know how to smile, and why reptiles don't. A new understanding of love takes form at the intersection of these disparate areas, wherein we can start to describe emotional life in a way true to known physiology and the life experience of human beings, their passions and anguish.

The scientist or the physician is not that terrain's sole surveyor, and certainly not its first. The aspiration to distill and transmit the secrets of the heart can attain a moment of matchless lucidity within a novel, a play, a short story, a poem. Through a symmetry as compact and surprising as the equivalence between matter and energy, love's poetry and its science share an unexpected identity. Each avenue uses the tools of the intellect to reach beyond; each seeks to lay hold of the ineffable and render it *known*, with the warm shock of recognition that truth so often carries. Now that science has traveled into the realm of the poetic, the efforts of one endeavor can inform those of its twin.

Long before science existed, sharp-eyed men and women told each other stories about how people are, stories that have never lost their power to enchant and instruct. The purpose of using science to investigate human nature is not to replace those stories but to augment and deepen them. Robert Frost once wrote that too many poets delude themselves by thinking the mind is dangerous and must be left out. That principle is mirrored in the study of the brain, where too many experts, out of plain fear, avoid mentioning love.

We think the heart is dangerous and must be left in. The poetic and the veridical, the proven and the unprovable, the heart and the brain—like charged particles of opposing polarity—exert their pulls in different directions. Where they are brought together the result is incandescence.

Within that place of radiant intersection, love begins to reveal itself. The journey we embark on here is by no means complete: the science of our day hints at structures but cannot define them. The castle of the emotional mind is not yet grounded in fact, and there is ample room left within its domain for conjecture, invention, and poetry. As neuroscience unlocks the secrets of the brain, startling insights into the nature of love become possible. That is what this book is about—and if that's not the secret of life, then we don't know what is.

Two

KITS, CATS, SACKS, AND UNCERTAINTY

HOW THE BRAIN'S BASIC STRUCTURE
POSES PROBLEMS FOR LOVE

Love fits with gliding ease into the heart of a troubadour's croon or a poet's couplet. There, in the mental balance weighing such correspondences, love indisputably belongs. But the prospect of putting humanity's palpitating heart under the scientist's steely gaze gives pause. Science operates under a bare but effective dictum: to understand a portion of the natural world, take it apart. Love is irreducible. The impasse looks definitive. How can investigation proceed? What can hard-edged objectivity apprehend about evanescent, ephemeral, *personal* love?

As it happens, science is less inimical to phantoms than it once was. The first years of the twentieth century crushed the conception of the natural world as a neat meshing of cogs, the inner details revealing themselves to any observer with a magnifying monocle of sufficient power and delicacy. As physicists and mathematicians delved deeper into the stuff of reality, they collided with the end of objectivity's jurisdiction. "O body swayed to music, O brightening glance / How can we know the dancer from the dance?" asked William Butler Yeats in 1928. The poet was in perfect harmony with the science of his age, which was reeling at the impossibility of dividing—as traditional science demanded— the knower from the known. Those hard-won lessons in scientific subjectivity can help us to understand why our age is at last on the brink of a revolution in humanity's vision of its own heart.

The first blow to the clockwork universe came from Albert Einstein. His relativity theory proposed that the flow of time depends on where you are, and that different observers may not agree even about the chronological order of the events they witness. A few years later, Kurt Gödel demonstrated that any mathematical system contains, like the gleaming and inaccessible jewels of a dragon's lair, true theorems that can never be proven. Between Einstein and Gödel came Werner Heisenberg and his uncertainty principle. Heisenberg showed that the more precisely one determines the position of an atomic particle, the less one can know about its speed. These shy qualities reverse their roles: the more exactly a particle's velocity is measured, the more elusive its location becomes.

The significance of Heisenberg's discovery expanded beyond the atomic level and recast the foundations of scientific endeavor. "Science does not describe and explain nature," Heisenberg concluded, but "nature as exposed to our method of questioning." Together with Gödel and Einstein, he introduced scientists to an uncomfortably indefinite world—where the extent of the knowable disappointingly dwindles, and such intangibles as *point of view* and *method of questioning* permeate previously solid truths. After 1930, mystery formed not only the perimeter of scientific knowledge but also its ineradicable center. For science to penetrate the mystery of love, its own style of questioning had to improve.

An old riddle illustrates how questions delimit the discoverable:

As I was going to St. Ives
I met a man with seven wives
Every wife had seven sacks
Every sack had seven cats
Every cat had seven kits
Kits, cats, sacks, and wives,
How many were going to St. Ives?

Many children know the best answer is one: the narrator alone is *known* to be bound for St. Ives. The listener isn't told and cannot divine the destination of the other travelers. The puzzle is fashioned to conceal the gap in the listener's knowledge. *How many were going to St. Ives?* yields an answer only by sweeping past the question crouching behind it—*Where is everyone going?* That question is unanswerable—and so it is rendered unthinkable. The cascading sevens distract the unwary as deftly as a conjurer diverts attention from a palmed ace. Seduced into the certain and the known, the listener is left scribbling away at an irrelevant calculation.

We cannot hope to unravel the heart's enigmas without knowing something about what love is made of, and how it operates. Biology has played almost no role in the most popular and influential views of love to date—because, as the St. Ives riddle portrays and Heisenberg proved, the questions we ask change the world we see. *What can the structure and design of the brain tell us about the nature of love?* could not have been glimpsed a hundred years ago. Absence of knowledge about the brain was not then deemed an impediment to understanding emotional life. Indeed, the omission was scarcely noticed.

Today, the relevance of love's physiology is here to stay. Love itself has not surrendered to reductionism, but in the last two decades of the twentieth century, the brain that produces love *did.* The advent of modern neuroscience, with its high-tech scanners and miniature tools of painstaking dissection, finally provided what the study of love had always lacked: a physical substrate that can be taken apart.

Seekers of the heart's secrets might be tempted to detour around the essential facts of brain structure, fearing the subject is impossibly technical and probably soporific. It is not. No one disputes that the brain's dense, delicate, filamentous intricacy inspires awe, and more than occasionally dismay. Those who wish to drink in the details, however, need not drown in them. Anybody can operate a car without an engineering degree. A working knowledge of internal combustion—what gasoline is, where it goes, and why you shouldn't peer into the tank with a lit match—is indispensable. You don't have to wade through back issues of *Scientific American* to grasp the nature of love, but acquaintance with the basics of the brain's origins and mechanisms can prevent some explosive misconceptions as passion's sparks begin to fly.

THE INSIDE STORY

The brain is a network of *neurons,* the individual cells of the nervous system. This account renders the brain in essence no different from the heart or the liver—organs that are also linked collections of similar cells. What gives an organ identity and power is the specialized function that its constituent cells stand ready to perform. The peculiar calling of a neuron is *cell-to-cell signaling.* Those signals are both electrical and chemical; the molecules whose restless shuttling sends the chemical portion of the message are the *neurotransmitters.* When people say that someone is afflicted with a "chemical imbalance" (now synonymous with "undesirable behavior beyond voluntary control"), they refer to one half of the signaling process, an unintended slight to a neuron's electrical potency. While few have witnessed electricity's flair for altering minds, everybody has seen chemicals change people. Coffee boosts alertness, alcohol dissolves inhibitions, LSD provokes hallucinations, and Prozac alleviates depression, obsessions, and low self-confidence—all by enhancing or disrupting these signals. Any substance that mimics or blocks native neurotransmitters can fiddle with an aspect of the mind: vision, memory, thought, pain, consciousness, emotionality, and yes, love.

What is the purpose of this assemblage of cells ceaselessly signaling one another? What useful property emerges from this festival of communication, and what end does it advance? Survival. A collection of signaling cells can engineer sudden reactions to instantaneous changes. Information from the environment can be translated into inbound signals, and after a flurry of internal processing within a centralized group of neurons, outbound signals produce *action:* a swipe at a fleeing morsel of food, or a leap to evade the pounce of a predator. Equipped with the best neurons firing in the best order, animals live longer. If they make it to the next mating season, they win. Natural selection awards no prize for second place.

Proud as we are of the nervous systems that tingle within our skulls, we should recognize that such an approach to the game of life represents one survival strategy among many. The world's most successful life-forms have no brains and no use for them. Bacteria, easily the most numerous creatures on Earth, are simple single cells that triumphed and persisted without any multicellular cooperative signaling or the complex behaviors that such communication bestows. Despite this seeming disability, they have exploited every ecological niche, from the arctic tundra to simmering sulfur hot springs. And the planet's longest-lived organism—the giant redwood tree of northern California, with a span of four thousand years—lives every minute of its nearly interminable life without the ability to react quickly to anything.

The earliest aggregations of signaling cells were sparse assemblies that embodied instructions for meeting the simplest environmental contingencies: encounter a noxious stimulus on the left, move right, and vice versa. Ages later, one hundred billion neurons make up the human brain. The brain's Byzantine conformation determines everything about human nature—including the nature of love.

The Triune Brain

No concerted development scheme forged the human brain. Evolution is a wandering process wherein multiple simultaneous influ-

ences, including chance and circumstance, shape biological structures over eons. A more capricious designer than any committee,
evolution is a story full of starts, setbacks, compromises, and blind
alleys, as generations of organisms adapt to fluctuating conditions.
We are accustomed to thinking of these adaptations as gradual and
progressive, but, as Niles Eldredge and Stephen Jay Gould argued
twenty-five years ago, the fossil record belies this impression.
Rather than a series of smooth transitions, the evolutionary
process is punctuated with bursts of metamorphosis. If an environment shifts fast enough or a favorable mutation arises, organismic modifications can explode into being.

Thus the development of the human brain was neither planned
nor seamlessly executed. It simply *happened*—and that pedigree nul-
lifies reasonable expectations about the brain's configuration. *A priori*, no one would suppose that advanced neural design should
require an organism to slip regularly into a helpless torpor that invites predation. But sleep is universal throughout the mammalian
world, although its neural function remains unknown. The same
fallible common sense suggests that the human brain is likely to be
unitary and harmonious. It isn't. A homogeneous brain might
function better, but humans don't have one. Evolved structures answer not to the rules of logic but only to the exigencies of their
long chain of survival victories.

Dr. Paul MacLean, an evolutionary neuroanatomist and senior
research scientist at the National Institute of Mental Health, has
argued that the human brain is comprised of three distinct sub-
brains, each the product of a separate age in evolutionary history.
The trio intermingles and communicates, but some information is
inevitably lost in translation because the subunits differ in their
functions, properties, and even their chemistries. His neuroevolu-
tionary finding of a three-in-one, or *triune*, brain can help explain
how some of love's anarchy arises from ancient history.

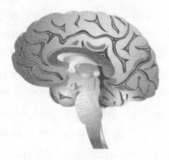

The human brain.

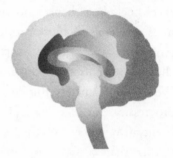

The triune brain.

The Reptilian Brain

The oldest or *reptilian* brain is a bulbous elaboration of the spinal cord. This brain houses vital control centers—neurons that prompt breathing, swallowing, and heartbeat, and the visual tracking system a frog relies on to snap a dancing dragonfly out of the air. The startle center is here, too, because a swift reaction to abrupt movement or noise is the principal reason animals have brains at all.

Steeped in the physiology of survival, the reptilian brain is the one still functioning in a person who is "brain-dead." If the reptilian brain dies, the rest of the body will follow; the other two brains are less essential to the neurology of sustaining life. Consider the railroad worker turned neurologic legend Phineas Gage. In 1848, an explosion drove a steel bar through Gage's skull; the rod entered below his left eye and exited the top of his head, taking a sizable piece of his neocortical brain and his reasoning faculty with it. Gage was a changed man after the accident, his diligence and tidiness forever transformed into sloth and disorganization. But after the blast, from the minute he sat up, Gage could walk and talk normally; he could eat, sleep, breathe, run, and gargle as fluently as any man. He lived another thirteen years without that cylinder of neocortical brain. Had the blast sent a spike hurtling through Gage's

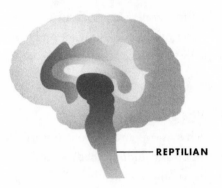

— **REPTILIAN**

The reptilian brain.

reptilian brain, he would have died before the first drops of blood hit the ground.

As long as the reptilian brain survives, it will keep the heart beating, the lungs expanding and relaxing, salt and water balanced in the blood. Like programmed appliances in a house whose owners have departed, a reptilian brain can plug away for years, despite the death of what makes a brain *human.* Our society greets with perplexity someone whose sole viable brain is reptilian: is such a person dead or alive? Is this a *person?* Sad as it may be, a body animated by the reptilian brain is no more human than a severed toe. The qualities that set us apart from other animals, or that distinguish one person from another, do not belong to this archaic conglomeration of cells.

We will be disappointed if we expect the reptilian brain to play a major role in the structure of the emotional mind. Reptiles don't have an emotional life. The reptilian brain permits rudimentary interactions: displays of aggression and courtship, mating and territorial defense. As MacLean notes, some lizard species attack and repel intruders from the district they have claimed as their own, illustrating just how primitive turf battles are in the history of terrestrial vertebrates. When we see urban gangs mark their domains and harass someone for stepping onto the wrong city block, or for

wearing a blue shirt in a zone where red shirts rule, we are witnessing, in part, a product of this antediluvian brain, with motivations
more suited to the lives of the asocial carnivores that brain was designed to serve.

THE LIMBIC BRAIN

In 1879, the French surgeon and neuroanatomist Paul Broca published his most important finding: that the brains of all mammals
hold a structure in common, which he called the *great limbic lobe* ("*le
grand lobe limbique*"). Because he could see a "line of demarcation"
between this convolution and the rest of the cerebral hemisphere,
Broca coined his term from the Latin word *limbus,* meaning "edge,
margin, or border." Since the structure he discovered marks the
evolutionary division between two disparate ways of life, his initial
designation proved unusually apt.

Humanity's second or *limbic* brain drapes itself around the first
with a languid ease. Within its smooth curves, however, lies a company of neural gadgets with tongue-twisting appellations. The limbic list sounds like the incantation of a magus: hippocampus,
fornix, amygdala, septum, cingulate gyrus, perirhinal and perihip-
pocampal regions.

Early mammals evolved from small, lizardish reptiles. The peculiar mammalian innovation—carrying developing young within a
warm-blooded body rather than leaving them outside in eggs—
had been established well before an errant asteroid rammed the
planet and put the chill on the dinosaurs. The rapid demise of the
reptilian giants left open opportunities for an upwardly mobile
class. Mammals scurried into the gap and bred like the rabbits they
were to become. Sixty-five million years later, the Age of Mammals
is still in full swing.

High school biology draws the distinction between reptile and
mammal along somatic lines: mammals sprout hair rather than

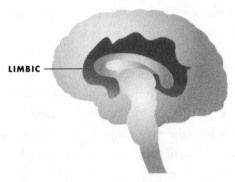

LIMBIC

The limbic brain.

scales; they are self-heating, while reptiles rely on the sun to regulate body temperature; they give birth to babies, not eggs. But MacLean pointed out that this classification overlooks a major brain difference. As mammals split off from the reptilian line, a fresh neural structure blossomed within their skulls. This brand-new brain transformed not just the mechanics of reproduction but also the organismic *orientation* toward offspring. Detachment and disinterest mark the parental attitude of the typical reptile, while mammals can enter into subtle and elaborate interactions with their young.

Mammals bear their young live; they nurse, defend, and rear them while they are immature. Mammals, in other words, *take care of their own*. Rearing and caretaking are so familiar to humans that we are apt to take them for granted, but these capacities were once novel—a revolution in social evolution. The most common reaction a reptile has to its young is indifference; it lays its eggs and walks (or slithers) away. Mammals form close-knit, mutually nurturant social groups—families—in which members spend time touching and caring for one another. Parents nourish and safeguard their young, and each other, from the hostile world outside their group. A mammal will risk and sometimes lose its life to protect a child or mate from attack. A garter snake or a salamander watches the death of its kin with an unblinking eye.

The limbic brain also permits mammals to sing to their children. Vocal communication between a mammal and offspring is universal. Remove a mother from her litter of kittens or puppies and they begin an incessant yowling—the *separation cry*—whose shrill distress drills into the ear of any normal human being. But take a baby Komodo dragon away from its scaly progenitor, and it stays quiet. Immature Komodos do not broadcast their presence because Komodo adults are avid cannibals. A lifesaving vacuum of silence stretches between a reptilian mother and young. Advertising vulnerability makes sense only for those animals whose brains can conceive of a parental protector.

And mammals can *play* with one another, an activity unique to animals possessing limbic hardware. Anyone who has joined a dog in a tug-of-war over an old sneaker, and has let the shoe go, knows what follows—he trots back. Mutual *tugging* is what he desires, not the shoe. The same dog appreciates the essential delight of keep-away played with a sock (doesn't want to keep the sock), and his heart warms to go-fetch—the improbably joyous celebration of making an object go exactly nowhere. What in the world do activities like this accomplish? The dog isn't finding food, isn't mating, isn't rearing pups, and isn't doing anything obviously linked to survival or propagation. So why do all kinds of mammals want to frolic, gambol, tumble, and roughhouse? For a mute mammal, play is physical poetry: it provides the permissible way, as Robert Frost said poems do, of saying one thing and meaning another. By the grace of their limbic brains, mammals find such exultant metaphor irresistible.

The Newest Brain

The *neocortex* (from the Greek for "new" and the Latin for "rind," or "bark") is the last and, in humans, the largest of the three brains.

Mammals that evolved long ago, like the opossum (so old that it retained the marsupial's trademark pouch), have only a thin skin of neocortex covering the older sub-brains. Neocortical size has grown in mammals of recent origin, so that dogs and cats have more, and monkeys, more still. In human beings, the neocortex has ballooned to massive proportions.

The human neocortex is two symmetrical sheets, each the size of a large, thick linen napkin, and each crumpled for better cramming into the small oblate shell of the skull. Like most of the brain, the neocortex is a warehouse of secrets and unanswered questions. Nevertheless, science has made some progress at mapping the functions and capacities of this massed neuronal army. Speaking, writing, planning, and reasoning all originate in the neocortex. So do the experience of our senses, what we know as awareness, and our conscious motor control, what we know as will.

The neocortical orchestration of our experiential world sometimes leads to surprising disjunctures of consciousness, the optical illusions of the self. Damage to the visual neocortex can produce the phenomenon of *blindsight,* wherein a patient reveals the erroneous impression of his own sightlessness. Although the world appears to his sensibility as a uniform and ceaseless night, if he is

NEOCORTICAL —

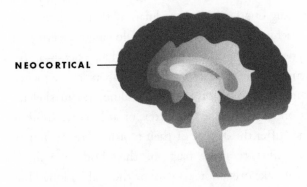

The neocortical brain.

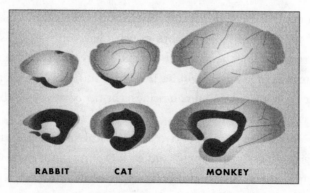

The brains of a rabbit, cat, and monkey. The neocortex has expanded in mammals of recent origin, while the size of the limbic brain has changed little. (From *The Triune Brain in Evolution* by Paul D. MacLean, 1990. Reprinted with permission of Plenum Press.)

forced to guess the location of a moving light he is correct far more often than chance permits—confirming to the external observer occult powers of vision that remain forever hidden from the patient himself. The delightful chronicles of Oliver Sacks include not only the man who mistook his wife for a hat, but also the man who mistook his own leg for a disembodied horror, and the woman who departed the bilateral universe by forgetting the concept of "left." All are examples of neocortical processing gone astray.

As a human being moves through the day, he is blissfully unaware of the prodigious feats of coordination that underlie the simplest acts. Reaching for a cup of coffee, allowing a greeting to roll off the tongue, glancing up Fifth Avenue to hail a taxi—all require the shortening of millions of tiny muscle fibers in a sequence of exquisite complexity. The cascade that culminates in skeletal muscle contraction begins in the neocortex, or at least we think it does. People who suffer the demise of their motor cortex (often by stroke) lose the ability to move parts of their body at will. If nearby neurons can take over some control of those abandoned but otherwise healthy muscles, a person may regain a limited capacity

to command them. The motor cortex thus emerges as a c
didate for the seat of volition.

Tracing the initiation of movement further back into the
undergrowth of the neural jungle soon reveals the brain's propensity
for frustrating such facile conceptions as a neat locus of control.
Recordings of encephalographic electrical waves show, amid their
jagged spikes and hieroglyph swirls, a signature downward dip sig-
nifying that a neuronal mandate for motion is under way: the so-
called *readiness wave*. While the motor cortex produces motion, the
readiness wave appears to signal intent. So we should look here for
will. When experimenters placed their subjects in front of a clock,
however, they found that the conscious experience of a decision to
move occurs *after* the readiness wave has already passed. What we
feel as the conscious spark of resolve in this case proves to be an af-
terthought, not the majestic nexus of initiative we might imagine.
Just where and how the early glimmers of intention coalesce, like
glittering dust motes into the swirling jinni of action, remains beyond the ken of today's science. The more we discover, the more we
find that we do not know. As E. E. Cummings observed: Always a
more beautiful answer that asks a more beautiful question.

While the neocortex may not supply a simple push button for
free will, small neocortical lesions can produce specific control
deficits—the inability to move an arm, to speak, even to focus attention. The functions of the older brains are involuntary. The
modulation of the blood's sodium concentration by the reptilian
brain, for example, occurs without a whisper of intent. So does the
startle response to a big bang—even with ample and detailed
warning, nobody can suppress flinching at a loud noise.

Another gift that the neocortex bestows is the skill of *abstraction:*
every task that calls for symbolic representation, strategy, planning,
or problem-solving has its headquarters in the neocortical brain.
That geography engenders the close relationship between neocor-

tex and conventional intelligence. Provided one corrects for body weight, species that are better problem solvers always have more neocortex than their less ingenious fellows. Human beings have the largest neocortex-to-brain ratio of any creature, an inequitable proportion that confers upon us our capacity to reason. Capacious neocortical abstraction also underlies the uniquely human gift of spoken and written language, in which meaningless squawks and squiggles stand for real people, objects, and actions. Language is the grandest and perhaps the most useful abstraction we have.

The power to symbolize arose not to grant the gift of gab but because it can keep an animal alive. Abstraction invents the possibility of a mental future. Because it can travel into the realm of the hypothetical, the neocortical brain can envision where and how a plan ends, allowing its possessor to strategize—rehearse and refine without betraying his intention prematurely, thereby allowing fictive mistakes whose corporeal counterparts he could not afford. The neurophysiologist W. H. Calvin has proposed that the cerebral neocortex originally developed to serve ballistic movements— complicated, one-shot actions that occur too rapidly to be modified as they uncoil, requiring planned precision. The modern Homo sapiens on the verge of shooting a crumpled paper ball into a distant wastebasket or lobbing keys to an acquaintance may experience today that moment of imaginative hesitancy before release, the preliminary, practice demitoss that sharpens aim. A talent for visualizing what-ifs may better someone's rock-throwing as much as his skill at chess. The former aptitude is what secured for the neocortex a lasting place.

Many people conceive of evolution as an upward staircase, an unfolding sequence that produces ever more advanced organisms. From this perspective, the advantages of the neocortex—speech, reason, abstraction—would naturally be judged the highest attributes of human nature. But the vertical conceptualization of evolution is fallacious. Evolution is a kaleidoscope, not a pyramid: tı. shapes and variety of species are constantly shifting, but there is no basis for assigning supremacy, no pinnacle toward which the system is moving. Five hundred million years ago, every species was either adapted to that world or changing to become so. The same is true today. We are free to label ourselves the end product of evolution not because it is so, but because we exist *now*. Expunge this temperocentrist bias, and the neocortical brain is not the most advanced of the three, but simply the most recent.

THE TROUBLE WITH TRIPLES

Evolution's stuttering process has fashioned a brain that is fragmented and inharmonious, and to some degree composed of players with competing interests. Critics of MacLean's triune model have disparaged its deliberate separation of intellect and emotion as unfashionable Romanticism. While the three brains differ in lineage and function, however, no one has argued for neurological autonomy. Each brain has evolved to interdigitate with its cranial cohabitants, and the lines between them, like dusk and dawn, are more shaded transitions than surgical demarcations. But it is one thing to say that night gives way to day and day fades into night, and it is quite another to declare light and dark equivalent. The cleavage between reason and passion is an ancient theme but no anachronism; it has endured because it speaks to the deep human experience of a divided mind.

The scientific basis for separating neocortical from limbic brain matter rests on solid neuroanatomical, cellular, and empirical grounds. As viewed through the microscope, limbic areas exhibit a far more primitive cellular organization than their neocortical counterparts. Certain radiographic dyes selectively stain limbic structures, thus painting the molecular dissimilarity between the two

n, vivid strokes. One researcher made an antibody that
ls of the hippocampus—a limbic component—and
those same fluorescent markers stuck to *all* parts of the
n, lighting it up like a biological Christmas tree, without
coloring tne neocortex at all. Large doses of some medications destroy limbic tissue while leaving the neocortex unscathed, a sharpshooting feat enabled by evolutionary divergence in the chemical composition of limbic and neocortical cell membranes.

Nor is there much room for doubt that nurturance, social communion, communication, and play have their home in limbic territory. Remove a mother hamster's whole neocortex and she can still raise her pups, but even slight limbic damage devastates her maternal abilities. Limbic lesions in monkeys can obliterate the entire awareness of others. After a limbic lobotomy, one impaired monkey stepped on his outraged peers as if treading on a log or a rock, and took food out of their hands with the nonchalance of one oblivious to their very existence. MacLean replicated the same loss of social faculties in rodents. After limbic ablation, adult hamsters ignored the calls and cries of their young; a limbectomized pup would repeatedly walk on top of the others "as though they did not exist." In addition to erasing the recognition of others, removing limbic tissue robbed these mammals of responsiveness to the playful overtures of normal littermates.

In humans, the neocortical capacity for thought can easily obscure other, more occult mental activities. Indeed, the blazing obviousness of cogitation opens the way to a pancognitive fallacy: I think, therefore everything I am is *thinking*. But in the words of a neocortical brain as mighty as Einstein's: "We should take care not to make the intellect our god; it has, of course, powerful muscles, but no personality. It cannot lead; it can only serve."

The swirling interactions of humanity's three brains, like the shuttling of cups in a shell game, deftly disguise the rules of emotional life and the nature of love. Because people are most aware of the verbal, rational part of their brains, they assume that every part of their mind should be amenable to the pressure of argument and will. Not so. Words, good ideas, and logic mean nothing to at least two brains out of three. Much of one's mind does not take orders. "From modern neuroanatomy," writes a pair of neuroscience researchers, "it is apparent that the entire neocortex of humans continues to be regulated by the paralimbic regions from which it evolved." The novelist Gene Wolfe makes an identical, albeit lovelier, observation:

> We say, "I will," and "I will not," and imagine ourselves (though we obey the orders of some prosaic person every day) our own masters, when the truth is that our masters are sleeping. One wakes within us, and we are ridden like beasts, though the rider is but some hitherto unguessed part of ourselves.

The scientist and artist both speak to the turmoil that comes from having a triune brain. A person cannot direct his emotional life in the way he bids his motor system to reach for a cup. He cannot will himself to *want* the right thing, or to *love* the right person, or to *be* happy after a disappointment, or even to be happy in happy times. People lack this capacity not through a deficiency of discipline but because the jurisdiction of will is limited to the latest brain and to those functions within its purview. Emotional life can be influenced, but it cannot be commanded. Our society's love affair with mechanical devices that respond at a button-touch ill prepares us to deal with the unruly organic mind that dwells within. Anything that does not comply must be broken or poorly designed, people now suppose, including their hearts.

Only the latest of the three brains traffics in logic and reason, and it alone can utilize the abstract symbols we know as words. The emotional brain, although inarticulate and unreasoning, can be expressive and intuitive. Like the art it is responsible for inspiring, the limbic brain can move us in ways beyond logic that have only the most inexact translations in a language the neocortex can comprehend.

The verbal rendition of emotional material thus demands a difficult transmutation. And so people must strain to force a strong feeling into the straitjacket of verbal expression. Often, as emotionality rises, so do sputtering, gesticulation, and mute frustration. Poetry, a bridge between the neocortical and limbic brains, is simultaneously improbable and powerful. Frost wrote that a poem "begins as a lump in the throat, a sense of wrong, a homesickness, a love sickness. It is never a thought to begin with."

Neither does love begin with a thought. Anatomical mismatch prevents intellectual talons from grasping love as surely as it foils a person who tries to eat soup with a fork. To understand love we must start with the feelings—and that is where the next chapter begins.

Three

ARCHIMEDES' PRINCIPLE

How we sense the inner world of other hearts

A body in water is subjected to an upward force equal to the weight of the water displaced. This is the skeleton of Archimedes' principle, true to mathematical relationships, cold to the touch. What breathes life into this dry dictum is the legend behind it. As the story goes, twenty-two centuries ago Hiero II, the king of Syracuse, commissioned Archimedes to determine if a certain crown was sterling gold or a tainted alloy. As Archimedes was stepping into his bath, he conceived of submersing the crown and comparing the amount of water it displaced to that displaced by an equal weight of solid gold. Any discrepancy between the two would indicate the crown and the test weight were different densities, and the crown, therefore, at least a partial fraud. This aquatic solution provided Archimedes with both his principle and its famous expression. After his inspiration, he is said to have run from his bath naked into the streets of the city, shouting, "Ευρηκα!" *

The centerpiece of this tale is not the crown or the gold or the cleverness, but Archimedes' passion, hot and pure. As Plutarch describes it:

> *Ofttimes Archimedes' servants got him against his will to the baths, to wash and anoint him, and yet being there, he would ever be drawing out of the geometrical figures, even in the very embers of the chimney. And while they were anointing of him with oils and sweet savours, with his fingers he drew lines upon his naked body, so far was he*

---

* In its English spelling, "Eureka!"

taken from himself, and brought into ecstasy or trance, with the de-
light he had in the study of geometry.

As elegant as his insight may be, it is the force of Archimedes' emotion that calls to us down the centuries. His thrill, not his intellectual dexterity, is what has given his theorem its notoriety. The real
principle behind his principle is that most people will never
fathom its mathematics—but his exuberance they do understand.
That rush of joy comes to some from seeing an out-of-the-park
home run, to others in the colors of the sun setting into the Pacific, or in the eyes of a newborn baby. Archimedes' delight transmits itself across two millennia in a heartbeat.

Why should we feel a kinship with Archimedes' enthusiasm, even
if his physics leave us tepid? To answer that question, we would first
have to know the answers to these: what *are* emotions? How do they
work? Where do they come from, and what are they for?

The superficial purposes of emotionality are plain. Exhilara-
tion, longing, grief, loyalty, fury, love—they are the opalescent pigments that gild our lives with vibrancy and meaning. And emotions
do more than color our sensory world; they are at the root of
everything we do, the unquenchable origin of every act more complicated than a reflex. Fascination, passion, and devotion draw us
toward compelling people and situations, while fear, shame, guilt,
and disgust repel us from others. Even the most desiccated neocortical abstractions pulse with an emotional core. Greed and ambition run beneath the surface of economics; vengefulness and
reverence under the veneer of justice. In all cases, emotions are humanity's motivator and its omnipresent guide.

Our society underplays the importance of emotions. Having allied itself with the neocortical brain, our culture promotes analysis
over intuition, logic above feeling. Cognition can yield riches, and
human intellect has made our lives easier in ways that range from

indoor plumbing to the Internet. But even as it reaps the
of reason, modern America plows emotions under—a cost
tice that obstructs happiness and misleads people about the
and significance of their lives.

That deliberate imbalance is more damaging than one might
suppose. Beyond the variegated sensations and the helpful motivations, science has discovered emotionality's deeper purpose: the
timeworn mechanisms of emotion allow two human beings to receive the contents of each other's minds. Emotion is the messenger
of love; it is the vehicle that carries every signal from one brimming
heart to another. For human beings, feeling deeply is synonymous
with being alive. In this chapter we will explore why.

THE SECRET SOCIETY OF MAMMALS

The first scientist to devote himself to the study of emotion was
Charles Darwin. After delivering *The Origin of Species*, Darwin wrote
three treatises that extended his ideas about evolution and natural
selection: *The Variations in Animals and Plants Under Domestication; The
Descent of Man, and Selection in Relation to Sex*; and *The Expression of the
Emotions in Man and Animals*, the last published in 1872. As his title
suggests, Darwin considered emotions an evolutionary adaptation
of organisms, no different from a host of other bodily modifications—claws, legs, stingers, gills, scales, wings. Natural selection
should favor emotionality for the same reason that it does any feature—enhanced survival. Organisms with an advantageous somatic
structure gain a competitive edge and live to pass their genes on to
the next generation, while those less equipped fade into the pale-
ontology texts. In Darwin's mind, emotions had to be bodily functions that persisted because of their inherent usefulness. He set
about dissecting emotional expressions to discern the underlying
biological utility he was certain they possessed.

After years of cataloguing emotional expressions as carefully as he did the bills of Galapagos finches, Darwin set out his conclusions. He proposed that the eyebrows lift in surprise to improve ocular mobility and the extent of the visual field, that the indrawn breath of a startle prepares one for a sudden flight that might follow, that the upturned lip of a socialite's sneer is the remnant of a dog's snarl, in which the animal exposes a cuspid to warn an opponent of its ferocity. Some of Darwin's hypotheses regarding the origins of expressions have scarcely been improved upon; others may strike us as fanciful. But however accurate his assertions about the profitability of individual expressions may have been, the essence of Darwin's approach was right on the mark. Emotions have a biological function—they *do* something for an animal that helps it to live, and if we study emotions carefully enough we might find out what.

Unfortunately, Darwin's evolutionary take on emotionality died an early death. As the study of the mind was launched at the beginning of the twentieth century, behaviorism soon dominated psychology, as psychoanalysis reigned over psychiatry. Both disciplines espoused views of emotion that were as distant from the evolution of terrestrial creatures as the moon. Darwin's ideas were relegated to obscurity for decades. For over fifty years, the preeminent theories of emotion in psychology and psychiatry were more philosophy than science: they were discussed and debated endlessly, tested rarely, and had only the faintest connection to human biology. In the mid-1960s, however, a handful of researchers revived Darwin's original concept of emotion as a heritable neural advantage. And the discoveries of the new emotion science have reshaped the modern vision of the mind, human nature, and love.

It's Not Just an Expression

Thirty years ago, emotion scientists Paul Ekman and Carroll Izard, working separately, confirmed a central proposition in Darwin's evolutionary theory of emotions: facial expressions are identical—all over the globe, in every culture and every human being ever studied. No society exists wherein people express anger with the corners of the mouth going up, and no person has ever lived who slits his eyes when surprised. An angry person appears angry to everyone worldwide, and likewise a happy person, and a disgusted one.

Convincing proof of universal emotional expressions came when Ekman reviewed 100,000 feet of movie film shot of isolated, preliterate tribes in New Guinea. The footage revealed that New Guineans make the same facial expressions as Americans. Despite differences in dress and appearance, in social milieu and custom, in climate and environment, and although none of them had ever seen a human being outside his own culture, the emotional expressions of the New Guinea natives were "totally familiar."

Ekman also tested their ability to recognize foreign facial expressions and found the same uniformity. He showed them three photographs of Americans—an angry, a happy, and a fearful face—and asked the natives to choose one that agreed with a story: "Her friends have come," or "She is about to fight." The New Guineans picked the glad photo for the former situation, the irate one for the latter. The natives were similarly adept at selecting American facial expressions that would match up with "Your child has died," and "You see a dead pig that has been lying there for a long time." Culture, Ekman found, doesn't determine the configuration of facial expressions: they are the universal language of humanity.

Proof that expressions are intrinsic is closer at hand than the South Pacific. As Darwin knew, a congenitally blind baby will smile while interacting pleasurably with his mother. Such a smile comes from a developing creature unable to speak, walk, or even sit up, but he already knows how to express happiness through a configuration of muscular contractions he has never seen on anyone's face. His knowledge has to be innate. A blind baby's smile must reflect the brain's inherited emotional architecture.

Ekman's work revealed that emotional expressiveness equips human beings with a sophisticated communications system. The receptive component allows people to acquire complex knowledge about the internal state of another person, irrespective of tribe or dialect. And all of us continually broadcast information about our inner states that any attentive human being can collect. Since emotions emanate from phylogenetic history, their antecedents must be found in other animals; our closest relatives should have emotional expressions that resemble ours. And they do.

NEUTRAL   ANGRY STARE   AFFECTION   EXTREME FEAR

Emotional expressions on the face of a rhesus monkey. (From Chevalier-Skolnikoff, 1973, in *Darwin and Facial Expression: A Century of Research in Review,* edited by P. Ekman. Reprinted with permission of the Academic Press.)

Because other mammals have expressions, does that mean they have *feelings*—a subjective experience of the emotional states they display? That idea was scientifically risible not so long ago. Now some emotion scientists endorse the proposition that other mammals possess emotional consciousness—that they *feel.* This reversal delights animal advocates eager to make an argument for panprotoplasmic parity. But when the zoophile Mark Derr writes, "The question of whether animals possess consciousness, intelligence, volition, and feelings has long been settled in the affirmative," he must be reporting the consensus from a species other than our own. Animals may have decided the matter to their own satisfaction, but human beings, as far as we know, are still debating it.

The case for animalistic subjectivity must rely on tangential evidence. We know, for instance, that some animals possess much of the same neural equipment that, in humans, gives rise to the experience of fear. If such an animal *looks* frightened (shows the facial expression of fear) and *acts* frightened (demonstrates the behaviors of fear, such as freezing, trembling, fleeing), then many reasonable people, including sober scientists, will conclude that it *feels* frightened.

Whether one is moved to endorse or reject the notion of zoological feelings, no proof can be adduced to narrow the gap. Subjectivity, by its nature, is nontransferable. (Even the supposition that other *people* feel rests beyond the perimeter of verifiability and, as we shall see, that commonplace assumption is occasionally incorrect.) Science holds marvels in store for future generations, but allowing human beings direct access to the inner sensations of a hedgehog or a dormouse will not be among them.

If we grant that the emotional club has a membership roster of more than one species, what other creatures should we nominate for inclusion? As a limbic product, emotionality belongs to the mammals. Snakes, lizards, turtles, and fish, lovable though they are to a select few humans, are not capable of perceiving or expressing emotional messages. They don't possess the requisite late-breaking brain.

An evolutionary hierarchy of emotion stretches from the first reptilian precursor to our own richly nuanced apparatus. Fear is probably the limbic brain's oldest emotion, an elaboration of the primordial reptilian startle. A touch of the heebie-jeebies helped early mammals safely navigate a world replete with dangers animate and inanimate—sharp teeth, dark caves, long claws, vertiginous heights. Disgust likewise serves to warn mammals of multifarious dangers—a tacit embodiment of Pasteur's germ theory of disease, disgust affirms the invisible likelihood of contamination from rotting foods and gelatinous excretions. Such lifesaving nauseous revulsion is at least as ancient as the animal that takes most advantage of its existence: *Mephitis mephitis*, the striped skunk.

The next emotions to sprout along the evolutionary tree assisted mammals in negotiating simple interactions. Anger readies a mammal for combat and warns others to expect a ferocious opponent. Jealousy alerts a mammal to the potential usurpation of reproductive chances. Later emotions inform social mammals with increasing precision about their status in a group—contempt, pride, guilt, shame, humiliation. The most recent emotions, and the ones least likely to be shared by other mammals, are those requiring a component of neocortical abstraction. Religious fervor is liable to be beyond the reach of nonhuman animals. So is the thrill that accompanies the realization of compact elegance in Pythagoras' theorem or Newton's gravitational law.

But most emotions require no thinking at all. For years, patients have told us stories about pets coming to their side and comforting them when they were distraught. Our medical training (often more hindrance than help in matters of the heart) led us to greet this allegation with a skeptical eye. How could a dog or a cat, with its diminutive brain, apprehend a phenomenon as complex as human emotion? One might as well expect an armadillo to master algebra. But cats and dogs are mammals—neocortically primitive and limbically mature. The limbic ancestry they share with humans should allow them to read and respond to certain emotional states of their owners. So when a person says he has a cat that can tell when he's had a bad day and hides under the bed, or a dog that detects sorrow and comes to console him, we no longer think he is being extravagantly anthropomorphic. The reciprocal process is dead easy: a perceptive human can tell if a dog is fatigued, contented, fearful, guilty, playful, hostile, or excited.

Not so with an animal that predates the limbic brain—try reading the inner state of a turtle, a goldfish, or an iguana. Animals with a common phylogenetic history share trait similarities: just as there are wide resemblances in the bony structure of the wrist or ankle among mammals, so too are there underlying commonalities in emotional perception and expression. Variants of the same emotional language exist throughout the mammalian family, some incomprehensible to us and others relatively close and accessible to our interpretative instrument, the limbic brain.

Music and Mayflies

Emotionality's code arises from a uniform neural architecture. The task of emotion science is to excavate this archaic structure, and as it has done so, it has unearthed the very roots of love.

Human beings, as tool-making animals, are prone to associate importance with durability. The columns of the Parthenon or the massive stone blocks of looming pyramids easily elicit our wonder and awe. The momentousness of emotions in human lives stands in befuddling contrast to their impossible brevity. Emotions are mental mayflies, rapidly spawned and dying almost as quickly as they arise. High-speed videography shows that facial expressions begin within milliseconds of a provocative event, and they fade immediately. We might sketch the concise life of a normal emotion in this way, with time extending along the horizontal axis and activity in the emotion circuits along the vertical:

Emotions possess the evanescence of a musical note. When a pianist strikes a key, a hammer collides with the matching string inside his instrument and sets it to vibrating at its characteristic frequency. As amplitude of vibration declines, the sound falls off and dies away. Emotions operate in an analogous way: an event touches a responsive key, an internal feeling-tone is sounded, and it soon dwindles into silence. (The figures of speech "pluck at one's heartstrings" and "strikes a chord in me" have found a home in our language for just this reason.) Rising activity in the emotion circuits produces not sound, but (among other things) a facial expression. When the neural excitation exceeds a shadowy threshold of awareness, what emerges is a *feeling*—the conscious experience of emotional activation. As neural activity diminishes, feeling intensity decreases, but some residual activity persists in those circuits after a feeling is no longer perceptible. Like the ghost of Hamlet's father, an emotion appears suddenly in the drama of our lives to nudge the players in the proper direction, and then dissolves into nothingness, leaving behind a vague impression of its former presence.

Moods exist because of the musical aspect of an emotion's neural activity, the lower portion imperceptible to our conscious ears. In our usage (adapted from Ekman), a mood is a state of enhanced readiness to experience a certain emotion. Where an emotion is a single note, clearly struck, hanging for a moment in the still air, a mood is the extended, nearly inaudible echo that follows. Consciousness registers a fading level of activation in the emotion circuits faintly or not at all. And so the provocative events of the day may leave us with emotional responsiveness waiting beneath our notice.

If a man spills coffee on himself, his annoyance is relatively short-lived—on the order of minutes. After the conscious feeling is gone, residual activity in the anger circuits lingers. He will pass into an *irritable mood*—a quickness to anger, the only reflection of the waning activity in those circuits. If he trips over his son's skateboard on the living room floor a bit later, his wrath will be faster and greater than the accident deserves on its own merits. Since the neural activation that creates a given emotion decreases gradually, provoking it again is easier within the window of the mood.

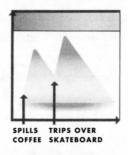

**SPILLS   TRIPS OVER
COFFEE   SKATEBOARD**

If emotions are ephemeral, how can we account for the person who feels sad all morning or frustrated all day? We must call upon the same generous pointillism that allows geometry to blend a collection of dimensionless dots into a uniform line or a graceful curve. The smooth impression of a lengthy emotion is often created by serial evocation, a repetitive string of one brief feeling that rings out its plangent tones again and again.

The most common precipitant of this reiterant emotionality is cognition: people tend to think about emotionally arousing occasions afterward, recirculating the experience and stimulating the consequent emotion just as if the inciting event had actually reoccurred. The human penchant for this post hoc cogitation can magnify the physiologic impact of an emotion many times. Anger sharply increases blood pressure on a short-term basis, for instance, but it may well be the recurrent stewing over provocative events that causes sustained hypertension in touchy people like type A executives. The neocortical brain's tendency to wax hypothetical then becomes a deadly liability. The limbic brain, unable to distinguish between incoming sensory experience and neocortical imaginings, revisits emotions upon a body that was not designed to withstand such a procession.

Certain brain configurations permit a single emotion to blare on unceasingly, without the rapid decay that typifies normality. Major depression is one such disease state, in which an acidic despair perpetually dominates the mind for weeks or months, sometimes blotting out all competing feelings, thoughts, and motivations. The manic extremity of bipolar disorder is another instance of uncommon durability in emotion, although in this case the irrepressible feelings tend toward euphoria and bonhomie. No one yet knows what causes the brain to get stuck on a single emotion, and in many cases, getting it unstuck is no simple matter.

AN EMOTIONAL EPIC

SCALES AND WIRES

Imagine a scene from 200 million years ago. A hatchling crocodilian rests motionless beneath the overhanging leaves of a damp fern, its mottled skin blending into the dirt and shadowed leaves. Jaws parted, tiny teeth bared, eyes unblinking, it might be carved from stone. On its left, a low-hanging branch yields a sibilant shudder as something large moves through the jungle. And with a push and paddle of its short legs, the young reptile splashes into the facing pond and disappears. For the moment, it has survived.

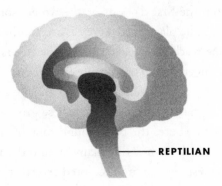

The reptilian brain.

Now let time spin through the intervening ages to the present. Continents fracture and slide across the globe, the ice caps extend and recede, countless species flash into existence and wink out again. But the crocodilian and the brain it possessed through all those millions of years remain essentially unchanged. The reptilian brain in our own skulls has not endured in that pristine state—it has adapted, changed, and learned to communicate with the two later brains that followed. Nevertheless, a rendition of the reptilian brain, the primal precursors of emotion within, is still contained within our own. The reptilian brain sits perched on the top of the human spinal cord, in appearance not unlike a bulbous frog crouching on a lily pad. Here one can find ancient control centers for vital bodily functions, including the primordial seeds of emotional responsiveness.

"Dream delivers us to dream, and there is no end to illusion. Life is a train of moods like a string of beads, and, as we pass through them, they prove to be many-colored lenses which paint the world their own hue. . . . Temperament is the iron wire on which the beads are strung." Writing these lines in 1844, Ralph Waldo Emerson may deserve credit as the first to propose that emotionality is hardwired. He was right: inborn emotionality is undeniable. From the first day out of the womb, some babies are criers, while others lie placid; some are easy to soothe and some inconsolable; some reach for a new rattle, while others shrink away. C. Robert Cloninger, M.D., has proposed that emotional control centers in the reptilian brain determine innate temperaments. Through the programmed responses of these groups of cells, the reptilian brain contributes the background tone to emotional life. That ancient brain becomes the filament upon which the later brains string the resplendent, multicolored crystals that merge into the mosaic of our emotional lives.

FRETFUL FRAMEWORK

Some people are risk-averse by nature: they save rather than spend, avoid rather than plunge, and hold back rather than let go. They have a temperament that tends toward *worry*, an aspect of emotional tone Cloninger thinks is controlled by the *raphe nucleus* in the reptilian brain. Worry is an inborn proneness to fear—an inclination to imagine future harm, and to activate the body's flight response system in case escape proves expedient.

The reptilian brain usually comes outfitted with a worry setting near the middle of the scale, a compromise that maximizes survival: too much fear is globally inhibiting, while too little promotes recklessness. The prehistoric crocodilian needed enough daring to venture out in the open from time to time, but it also required the wariness that allowed it to slip into the pond on a moment's notice. Most people have a moderate amount of inbred worry, although our popular culture is fond of idealizing individuals whose worry is nonexistent. Arnold Schwarzenegger and Bruce Willis are the latest in a string of actors whose screen personas wisecrack coolly in the face of heart-stopping danger. When we identify with their bravado, we treat ourselves to the vicarious thrill of a temperament most can never experience.

For this privation we should feel only gratitude. In the historical jungle of evolution, a minute level of worry invited disaster too often. Many of our ultralow-anxiety ancestors were bitten by snakes, gored by tusks, and fell out of trees. Those premature deaths shifted the gene pool toward higher trepidation. Children born today with a diminutive level of worry—those whose emotional physiology underreacts to stress, novelty, and threat—grow up to become criminals much more often than average. Criminality has long been known to be partially heritable, and a worry volume set to "low" in the reptilian brain is part of the mechanism. Anxiety deters people from high-risk acts. Those who do not experience the emotional weight of adverse consequences will not be sufficiently warned off. They will not know when they are about to do something they should by all rights fear and avoid.

As DNA shuffles and recombines in humanity's gene pool, the unlucky inherit extremes of temperament. For the most part, their eccentric dispositions will not serve them well. Before they dared to creep from beneath the protection of a fern, the reptilian precursor to worry gave our predecessors the hesitance to act and the predilection to flee that saved their lives. While the locus of danger in our lives has changed, the underlying neural mechanisms remain. Those worry circuits still perform the same function: under their direction, people imagine future harm, withdraw from potential threats, and their hearts, lungs, and sweat glands warm up for sudden use. An unfortunate few suffer from a hair-trigger sensitivity in this primordial system. When the neural alarm apparatus goes off with a bang, the result is a panic attack—a paroxysm of terror, an explosion of somatic sensations and reactions (chest tightness, racing heart, sweaty palms, churning stomach), and an outpouring of fear-soaked expectations and plans.

When anxiety becomes problematic, most people try vainly to think their way out of trouble. But worry has its roots in the reptilian brain, minimally responsive to will. As a wise psychoanalyst once remarked of the autonomic nervous system (which carries the outgoing fear messages from the reptilian brain), "It's so far from the head it doesn't even know there is a head." A high-worry temperament, however, does not doom its every possessor to a lifetime of anxiety. The brute force of will cannot undo temperament. But, as we shall see in later chapters, subtler means of emotional influence exist that can tame even the wild beast of panic.

The emotion circuits in the reptilian brain, like those responsible for worry, create a broad behavioral disposition. We can glimpse in them the earliest form of a neural system that scans the environment and quickly prepares an animal's physiology for the lifesaving response—as when a young reptile slides into the safety of a lagoon at the hint of a nearby predator. But the perceptive range of a reptile is limited, and the reptilian brain alone can orchestrate only coarse physiologic changes. With the arrival of the limbic brain, the neural resources aimed at coordinating physiology and environment expanded lavishly. When evolution brought mammals into being, it created an organism with a novel kind of neural responsiveness—one that permitted the intimate mental embrace of love.

The Bridge Between Worlds

In 1792, George Shaw of the Royal Zoological Society in London received a specimen from Australia. He found before him a squat, spiny creature, something like an undersized porcupine bearing a protuberant hollow snout. Shaw did not realize that he was holding a remnant of one of evolution's most important crossroads—the one that led to the birth of mammals.

The echidna, the creature Shaw received, is technically classified as a mammal, but it is either the most reptilian mammal or the most mammalian reptile imaginable. Echidnas locomote with the

The Australian echidna.

same low-slung, waddling gait as a lizard. They lead solitary lives, coming into each other's company only long enough to mate. And the product of that copulation is a leathery, reptilian egg, which the female carries next to her body between two elongated folds of skin—an open-air uterus. An egg-laying mammal was a bewildering conflation of reptile and mammal to the classifications of nineteenth-century science. Most experts of the day refused to believe that monotremes—the taxonomic category to which echidnas belong—were truly egg layers. The naturalist William Caldwell provided definitive proof in 1884, when he saw with his own eyes an egg in an echidna's primitive pouch. His telegram back to civilization—*"Monotremes oviparous"*—rocked the scientific world.

Arising somewhere between 100 million and 150 million years ago, monotremes demonstrate the beginnings of the departure from a reptilian way of life. Although it was far from obvious to early taxonomists, the feature that distinguishes mammals from reptiles is the appearance of a new brain within their skulls—the limbic brain. The echidna possesses not only nature's most primitive uterus, but also her most primitive limbic apparatus. Of all mammals, echidnas alone lack one limbic process: they do not dream during sleep.

In its present form, the limbic brain is not only the seat of dreams, but also the center of advanced emotionality. The primor-

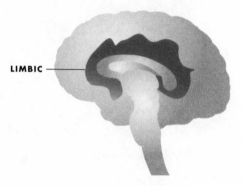

The limbic brain.

dial purpose of the limbic brain was to monitor the external world and the internal bodily environment, and to orchestrate their congruence. What one sees, hears, feels, and smells is fed into the limbic brain, and so is data about body temperature, blood pressure, heart rate, digestive processes, and scores of other somatic parameters. The limbic brain stands at the convergence of these two information streams; it coordinates them and fine-tunes physiology to prime the body for the outside world.

Some of these modulations are immediate, such as changes in sweating, breathing, or heart rate. The limbic brain effects these alterations through its connections to the control centers of the reptilian brain. Other bodily changes of limbic origin are longer-lasting: its outputs to the endocrine system allow emotional states to affect global bodily functions like immune regulation and metabolism. The neocortical brain, although a latecomer to the emotional scene, also receives limbic directives. These influence the tone of symbolic activities, like language, and strategic operations, like action planning. And the limbic brain orchestrates brain changes that serve a purely communicative role—in response to limbic stimulation, small muscles on the mammalian face contract in precise configurations. The face is the only place in the body where muscles connect directly to skin. The sole purpose of this arrangement is to enable the transmission of a flurry of expressive signals.

Consider, for instance, this situation: a man is riding to work on a bus, heading for the financial district in downtown San Francisco. A tattooed teenager with a shaven head (not a rarity in these parts) boards the vehicle, glares at the commuter, and bumps by him. That sensory experience flashes to the limbic brain, which will sift the event for its significance and prepare physiology to meet *that* singular moment. Our man's limbic brain will receive input about the intruder's facial expression, his pupil size, his body posture and gait, and perhaps even his scent. The limbic brain evaluates the nature of the other's intention—is it careless, aggressive, friendly, sexual, submissive, indifferent? A given limbic brain arrives at conclusions based on the collaboration of its genetically specified wiring scheme and past experience of similar situations. In this case, let us suppose our man's limbic apparatus detects hostility and, to meet the situation, equips him with the emotion of anger.

Once the limbic brain has settled on an emotional state, it sends outputs to the neocortical brain, spawning a conscious thought (*Who the hell does this guy think he is?*). At the same time, limbic outputs to the premotor areas of the neocortex are directing action-

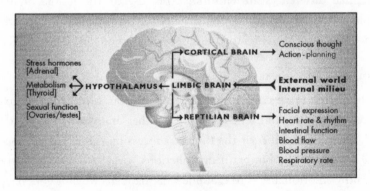

The centrality of the limbic brain.

planning. Meanwhile, outputs to the endocrine system will alter stress hormone release, which may impact the entire body for hours or days afterward. Limbic instructions to the lower brain centers will cause facial muscles to contract in the configuration of anger: eyes narrowed, brows drawn together, lips pressed, with the edges of the mouth turned down. The limbic brain will direct the reptilian brain to change cardiovascular function. Heart rate will increase, as will blood flow to the arms and hands—because the outcome of anger may be a fight, the limbic brain readies the physiologic systems most suited to fisticuffs. The entire maneuver is executed with the speed and grace of a ballerina's pirouette. One moment a man is minding his own business—two seconds later, anger swells, his brow furrows, and his hands start to clench.

Suppose that a woman follows just behind the belligerent youth. Witnessing the encounter, she shoots our traveler a look of sympathetic recognition and mock exasperation. *Can you believe what it's like on the buses these days?* she might say, if she were speaking. She isn't. But our commuter's limbic brain will nevertheless discern the message in her eyes and her face. To an emotionally insensate organism, the two interactions look exactly the same: for an instant, a moving person glanced at another. But the emotional implications of the infinitesimal differences are enormous. Because of the limbic brain's split-second precision, one can successfully distinguish an impending fight from the empathetic communication of kindred spirit.

SOLITARY CONFINEMENT

The limbic brain collects sensory information, filters it for emotional relevance, and sends outputs to other brain areas thousands of times a day. Most of the time its processing is flawless, but occasionally the limbic brain malfunctions. One way to appreciate healthy emotionality is to examine what happens when it goes haywire. Human beings are immersed in a sea of social interchange, surrounded by a subtle communications network that most do not notice. The limbic brain is our internal cryptographic device, allowing us to decipher a flood of complex messages in an instant. But when decoding breaks down, the resulting deficits can show us what emotionality enables the rest of us to do.

Several years ago we encountered a sixteen-year-old high school sophomore whom we'll call Evan. His mother wanted him to see a psychiatrist because she was concerned about his lack of friends. Other children had teased and rejected him since he was a young boy.

Upon meeting Evan, it was not hard to understand the derivation of the taunts. Evan was pleasant and friendly, but his social behavior was discordant and jarring. He stood too close when shaking hands, for instance, and he spoke too loudly. His voice was strangely flat, his eye contact sporadic, and his style of dress atypical for California teenagers: a plaid shirt with a solid blue tie.

Evan's professed purpose was not to attain eccentric prominence among his peers. He was genuinely confused by their rejection, and he wanted to know what he could do to get along with them better. His intellect was keen and his grades excellent, but as we got to know him better, we discovered that Evan was completely unable to intuit the rules of social interchange—hence his dress, manner, and style of greeting. He once tried to ask a girl out by presenting her with a lollipop. She thought he was making fun of her and became angry. He, in turn, was baffled by her reaction. As he explained, he had observed that people proffer gifts as a token of friendship, including the occasional lollipop.

Most of us understand that lovers exchange flowers, candy, and poems, while lollipops are given to children and birthday celebrants. Who can say why lollipops do not express romance? The code that governs this conduct is surely capricious, but most people have no trouble interpreting it. This boy didn't acquire social conventions naturally; even with monumental effort they persistently eluded him. He could accept concrete guidelines about human interchange, like "Most people expect you to stand about *this* far away when you speak to them." But he could not grasp the *essence* of interaction—he was never able to pick up on another's discomfort and adjust his distance accordingly, as a limbically fluent person would. Emotional signals remained obscure hieroglyphics to him. The limbic brain that should have given him the Rosetta stone to his emotional life had failed him. He remained lost, a socially blind person in a relentlessly social world.

The Viennese pediatrician Hans Asperger first described this affliction in the 1940s; it is now known as *Asperger's syndrome*. Children with Asperger's can be intellectually bright or brilliant, but they are emotionally clumsy, tone-deaf to social subtleties in others, and sometimes to their own emotions. When we asked a young woman with Asperger's what made her unhappy, she was quick to correct us: "I know that the words *happy* and *unhappy* signify something to other people, and I have heard others use them, but I do not know what they mean," she told us. "As far as I know, I have had no experience of either. I have no basis on which to answer your question." Startled, we tried to find a broader area of emotionality she could relate to. "Do you have a sense of what it's like to play?" one of us asked her. She stared for a moment, puzzled, and then asked, "As opposed to what?"

Finishing Touches

Because the last brain in the evolutionary sequence directs the abstract mind, we must credit the neocortex for the towering human achievements in cognition—language, problem-solving, physics, mathematics. Emotional function doesn't require many hypotheticals—it takes neocortical genius to formulate the theory of relativity, but not to be sad after a loss, or to be thrilled at seeing the person you love across a crowded room. But while the neocortical brain does not produce emotionality, it does have a role in modulating feelings and integrating them with some of its own symbolic functions.

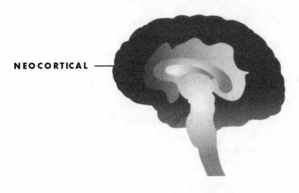

NEOCORTICAL

The neocortical brain.

IN A DIFFERENT VOICE

With its power to weave and unravel abstractions, the neocortex produces language—a string of arbitrary symbols that convey a message. While having emotions is under limbic control, *speaking* of them falls under the jurisdiction of the neocortex. That division of labor creates translation troubles. One of the neural mechanisms that bridges the gap is *prosody*—a process the neocortex borrows to inflect its dry concepts with emotional relevance.

The two language centers of the brain reside in the *left* temporal neocortex.

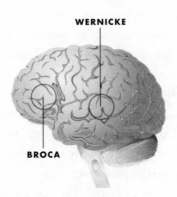

WERNICKE

BROCA

Language centers on the left side of the neocortical brain.

Wernicke's area translates the whistles and clicks of inbound speech into meaning, while *Broca's area* spins thoughts into a steady string of words. People with damage to Wernicke's area cannot understand what is said to them, though they can express themselves verbally, while those with damage to Broca's area can no longer talk, but they can still comprehend others who do.

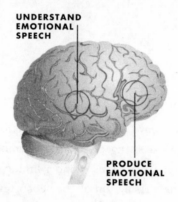

Emotional language centers on the right side of the neocortical brain.

The mirror-image areas of the *right* temporal neocortex perform the same functions on the emotional content of speech. People with damage to these areas evidence *aprosodia:* a significant fraction of them can no longer discern the emotional meaning of speech, while others cannot deliver emotional nuances in spoken language. These are crippling deficits, because sentences with identical semantic structure can easily have opposite meanings when they differ in prosody. Sarcasm owes the whole of its existence to tone. A sentence as apparently straightforward as "That's a nice haircut" is thoroughly ambiguous without prosody—the words can convey anything from "I'd like to go to bed with you" to "You look like a fool." Anyone who has cohabited with a teenager knows that single monosyllables—yeah, right, sure—can express assent, contempt, enthusiasm, indifference, or a thousand other delicately shaded meanings. A person with damage to the right-side mirror of Wernicke's area can't distinguish among the limitless possibilities mere words suggest. Those with damage to the dextro–Broca's area can't imbue their speech with emotional inflections—where they should be able to draw upon the chromatic palette of emotions to tinge their words with felt meaning, their speech remains dull and opaque. Their words cannot sound threatening, playful, or affectionate, which makes it nearly impossible to communicate successfully with emotionally fluent human beings, who rely on those clues to derive a speaker's intention.

While damage to the right temporal neocortex is fairly rare, millions of people experience daily aprosodia in their e-mail. At night all cats are gray, and in e-mail everyone is aprosodic, because the medium consists of curt sentences lacking emotional inflections. This is why people misunderstand one another so readily by e-mail, and why it is so much easier to lie on the Internet than in other social interactions. Minus the perceptible cues of voice tone, eye contact, and expression, e-mail so lends itself to emotional deception that people assume outrageously fabricated identities, simply because they can.

The human need for prosody is too great to go unanswered, and so it has spawned text-based emotional inflectors, *emoticons*. An emoticon sketches a facial expression with a couple of punctuation marks—to derive the meaning, the viewer mentally rotates the image ninety degrees clockwise. Pleasure and displeasure were the broad emotional states first so caricatured and communicated—

:)   :(

—and as the popularity of e-mail has exploded, so has the inventiveness of emotional iconographers. More than two hundred emoticons now exist for conveying a raft of mental states ranging from mischievous:

>:–)

to astonished:

#:–0

The quick rise of emoticons confirms the intolerable ambiguity of the neocortical brain's advanced symbolic tools, and the problem this poses for successful communication between limbic creatures. But no matter how creatively designed, emoticons cannot compete with emotions—a delicately decorated parenthesis cannot depict nostalgia, jealousy, wistfulness, or envy. In our increasingly digitized world, e-mail is a convenient substitute for dialogue, but it does not convey the richness that humans unthinkingly transmit when they use emotionally tempered speech and facial expressions.

That missing limbic data is extraordinarily valuable. Telecommunications giants are currently sinking hundreds of millions of dollars into the race to develop affordable two-way video sent over a phone line or a cable television connection. Even with advanced information compression algorithms, a data stream with resolution fine enough to catch the subtleties of facial expression requires about four hundred kilobits per second. That should give us an idea of the massive sensory fire hose the limbic brain is tapping into as it discriminates remorse from disdain, delight from terror, indignation from admiration.

A RESOUNDING SUCCESS

Animals with little neocortical brain—dogs, cats, opossums—have emotions. So does the world's most interesting noncognitive mammal, the human infant. Infants are early masters of detecting and expressing emotions, which may help to explain their inborn fascination for faces. If you want to capture the attention of an infant, you will have more luck using an expressive human face than any other object in the world. Babies have an intrinsic appetite for faces: they look at them, peer at them, gaze at them, stare at them. But what exactly are they looking *for*?

Researchers now know that babies are looking at the expressions on the faces they fix on. In studying what attracts infant attention, researchers rely on measurements of gaze, because babies will look longer at novel objects than familiar ones. One can demonstrate in this manner that infants just a few days old can distinguish between emotional expressions.

What is so important to a baby about knowing his mother's emotional state? A scenario called the *visual cliff* suggests an answer. A baby is placed on a countertop, half solid and half clear Plexiglas. From the baby's point of view, he reaches an abyss when the Plexiglas begins, and he seems in danger of falling. The translucent plastic provides real, albeit invisible support, and thus, the visual cliff presents babies with an ambiguous threat. To an infant unschooled in the nature of Plexiglas, it appears he will fall, but since the surface is solid to the touch, he can't be sure. How does he make sense of it?

A typical baby crawls to the edge of the cliff, sees the possible precipice, and then looks at his mother—and makes his assessment of the cliff's lethality by reading her expression. If she radiates calm, he continues crawling, but if he finds alarm on her face, the baby stops in his tracks and cries. Whether they realize it or not, mothers use the universal signals of emotion to teach their babies about the world. Because their display is inborn, emotions not only reach across the gaps between cultures and species, but they also span the developmental chasm between mother and infant. Emotionality gives the two of them a common language years before the infant will acquire speech, the arbitrary symbolic system of the neocortical brain.

But an infant doesn't check up on his mother's face only when ambiguity threatens—babies continuously monitor their mothers' expressions. If a mother freezes her face, her baby becomes upset and begins to cry in short order. How much expressiveness do babies demand? Imagine a double video camera setup, in which mother and baby can see each other, but not face-to-face; each sees the other in their respective monitors. In real time, mother and infant look at each other, smile and laugh, and both are perfectly happy. If the baby sees a videotape of his mother's face instead of the real-time display, he quickly becomes distraught. It isn't just his mother's beaming countenance but her *synchrony* that he requires— their mutually responsive interaction. Restore his mother's face in real time to his TV monitor, and his contentment returns. Introduce a delay into the video circuit, and the baby will again become distressed.

An infant can detect minute temporal changes in emotional responsiveness. This level of sophistication is coming from an organism that won't be able to stand up on his own for another six months. Why should a creature with relatively few skills be so monomaniacally focused on tiny muscular contractions visible beneath the skin of another creature's body?

The answer lies in the evolutionary history of the limbic brain. Animals have highly developed neural systems for processing specific informational needs. The sonar system of bats serves them admirably in chasing small bugs in a pitch-black night; within the cacophony of their high-pitched echoes, they can see a world we are blind to. The intricate cellular structure of certain eels allows the precise mapping of perturbations in nearby electric fields; the eel recognizes other fish, including its prey, by the pattern of electricity their muscles cast off.

The limbic brain is another delicate physical apparatus that specializes in detecting and analyzing just one part of the physical world—the internal state of other mammals. Emotionality is the social sense organ of limbic creatures. While vision lets us experience the reflected wavelengths of electromagnetic radiation, and hearing gives information about the pressure waves in the surrounding air, emotionality enables a mammal to sense the inner states and the motives of the mammals around him.

The reptile brain, capable of reading the world and altering internal physiology to meet changing conditions, contains the germ of emotion. In mammals, emotionality vaulted to a vastly more sophisticated level. A young crocodilian can sense a possible predator behind a wavering frond, and it can mobilize its physiology to evade the threat. But a mammal can turn its advanced neural sensor not only on the inanimate world but also on other animals that are emotionally responsive. A mammal can detect the internal state of another mammal and adjust its own physiology to match the situation—a change in turn sensed by the other, who likewise adjusts. While the neural responsivity of a reptile is an early, tinny note of emotion, mammals have a full-throated duet, a reciprocal interchange between two fluid, sensing, shifting brains.

Within the effulgence of their new brain, mammals developed a capacity we call *limbic resonance*—a symphony of mutual exchange and internal adaptation whereby two mammals become attuned to each other's inner states. It is limbic resonance that makes looking into the face of another emotionally responsive creature a multilayered experience. Instead of seeing a pair of eyes as two bespeckled buttons, when we look into the ocular portals to a limbic brain our vision goes deep: the sensations multiply, just as two mirrors placed in opposition create a shimmering ricochet of reflections whose depths recede into infinity. Eye contact, although it occurs over a gap of yards, is not a metaphor. When we meet the gaze of another, two nervous systems achieve a palpable and intimate apposition.

So familiar and expected is the neural attunement of limbic resonance that people find its absence disturbing. Scrutinize the eyes of a shark or a sunbathing salamander and you get back no answering echo, no flicker of recognition, nothing. The vacuity behind

those glances sends a chill down the mammalian spine. The prelim-
bic status of mythological creatures that kill with their gaze—
the serpent-crowned Medusa, the lizardlike basilisk, hatched from
a cock's egg by toads or snakes—is no accident. These stories create monsters from ordinary reptiles by crediting them with the
power to project out of their eyes what any mammal can see already dwells within: cold, inert matter, immune to the stirrings of
limbic life.

To the animals capable of bridging the gap between minds, limbic resonance is the door to communal connection. Limbic resonance supplies the wordless harmony we see everywhere but take
for granted—between mother and infant, between a boy and his
dog, between lovers holding hands across a restaurant table. This
silent reverberation between minds is so much a part of us that,
like the noiseless machinations of the kidney or the liver, it functions smoothly and continuously without our notice.

Because limbic states can leap between minds, feelings are con-
tagious, while notions are not. If one person germinates an ingenious idea, it's no surprise that those in the vicinity fail to develop
the same concept spontaneously. But the limbic activity of those
around us draws our emotions into almost immediate congruence.
That's why a movie viewed in a theater of thrilled fans is electrify-
ing, when its living room version disappoints—it's not the size of
the screen or the speakers (as the literal-minded home electronics
industry would have it)—it's the *crowd* that releases storytelling
magic, the essential, communal, multiplied wonder. The same limbic evocation sends waves of emotion rolling through a throng,
making scattered individuals into a unitary, panic-stricken herd or
hate-filled lynch mob.

It seems a strange irony that we need science to rekindle faith in
the ancient ability to read minds. That old skill, so much a part of
us, is not much believed in now. Those who spend their days with-

out an opportunity for quiet listening can pass a lifetime and over-look it altogether. The vocation of psychotherapy confers a few unexpected fringe benefits on its practitioners, and the following is one of them. It impels participation in a process that our modern world has all but forgotten: sitting in a room with another person for hours at a time with no purpose in mind but attending. As you do so, another world expands and comes alive to your senses—a world governed by forces that were old before humanity began.

Four

A FIERCER SEA

How relationships permeate the
human body, mind, and soul

When Romeo hears of Juliet's demise (a report that turns out to be false), he goes immediately to her mausoleum to join her in death. So mad with grief and intent on their funereal reunion is he that he says to Balthasar, his trusted servant, who he thinks may try to stop him:

> *But if thou, jealous, dost return to pry*
> *In what I further shall intend to do,*
> *By heaven, I will tear thee joint by joint,*
> *And strew this hungry churchyard with thy limbs.*
> *The time and my intents are savage-wild,*
> *More fierce and more inexorable far*
> *Than empty tigers or the roaring sea.*

Love is no less ferocious today. Romeo's anguished cry rings true because it resonates within the same emotional architecture the Bard intuited. What is the nature of aching loss and the desperate urge for reunion with those we love? What makes passion savage and inexorable? Our culture has forgotten that primordial knowledge, now buried beneath an impenetrable layer of lectures and instructional videotapes. Relationships have taken on the status of weather—everyone talks about them, but who knows what to *do*?

Relatedness, affiliation, loyalty, and nurturance are woven so thoroughly into our lives that we tend to presuppose their ubiquity throughout the animal kingdom. But most creatures do not know these motivations. Cannibalism—specifically, parents ingesting offspring for nutritive value—revolts human beings, but for many species the line between progeny and delicacy is blurred. A friend who kept guppies gave them up upon realizing she would have to segregate the young to prevent their wholesale consumption by parents. Such indiscriminate dining habits are, the pet store manager told her, normal for grown guppies. "Not in my house, they aren't," she answered grimly, and expelled the fish into the toilet. If they made it to the sea, they may be cannibalizing still. The diminutive crocodilian we encountered in the last chapter had good reason to be wary: nine out of ten baby crocodiles finish life in the belly of a predator before their first birthday; in most instances, the poacher is an adult croc. Given how primal the urge is to gobble up a smaller organism, feelings of tenderness, care, and concern for the tiny and frail may rightly strike us as near marvels. They are limbic endowments, and so are the rage and tears that erupt at the fracture of a mammalian bond. Of what is that miraculous tie made? For animals as social as we, that question defines our lives.

FINDING THE TIES THAT BIND

The Austrian physician and Nobel laureate Konrad Lorenz launched the scientific study of relatedness in response to a children's book. The child of parents "supremely tolerant of my inordinate love for animals," Lorenz grew up on a large estate in Altenburg, Germany, where he kept a menagerie of insects, fish, reptiles, dogs, and monkeys. But after he read *The Wonderful Adventures of Nils*, in which a mischievous boy joins a flock of migrating geese, Lorenz's avian pets became his lifelong love. "From then on, I yearned to become a wild goose and, on realizing that this was impossible, I desperately wanted to have one," Lorenz wrote. His devoted observations of the waterfowl in his backyard convinced him that much of their behavior, including mother-offspring bonding, was instinctual. Lorenz's best-known studies concerned ducklings and goslings, who huddle by their mother while she rests, and clamber after her when she is on the move.

Baby ducks tagging along behind their mother are a familiar sight to anyone acquainted with kindergarten reading primers. But how, Lorenz wondered, do they know whom to follow? As a boy, he was delighted to see that hatchlings would trail after him instead of their mothers. As a scientist, Lorenz found that ducklings would tail *anything*—no matter how implausible a mother—provided they saw it move early in their lives.

Lorenz realized that when goslings in the wild follow a mother goose, they do so not because they recognize a parent who will lead them to food and away from danger. Evolution has instead equipped goslings with a hardwired neural rule ("follow *that*"), and the dictum applies to any object falling within some sketchy guidelines for motherhood ("seen early in life" plus "moving"). The first entity a freshly hatched bird normally sees *is* its mother, but the bird's neural system is programmed to detect only a few of her relevant characteristics before fixating on her, and the system can be fooled. Lorenz used the word *imprinting* for the tendency of birds and mammals to lock on to an early object. In work done since, lambs have been tricked into forming a bond to television sets, guinea pigs to wooden blocks, and monkeys to cylinders of wire bent into the rough outline of a simian mother.

Imprinting is a manifestation of rudimentary neural systems dabbling in relatedness, and its rigidity owes much to the primitive nature of those circuits. Human relationships show similarly lawful properties. Even though primate attachments are more flexible than a gosling's, they bend much less than people expect.

Frederick II, a thirteenth-century Holy Roman emperor and

king of southern Italy, unwittingly conducted the first study of human bonding. His Imperial Majesty, who spoke several languages himself, thought he could determine the inborn language of mankind by raising a group of children who would never hear speech. Saltimbene de Parma, a Franciscan monk who chronicled the exploits of the experimental monarch, wrote that Frederick proceeded by "bidding foster-mothers and nurses to suckle and bathe and wash the children, but in no wise to prattle or speak with them; for he would have learnt whether they would speak the Hebrew language (which had been the first), or Greek, or Latin, or Arabic, or perchance the tongue of their parents of whom they had been born." But, the good brother wrote, Frederick's exercise terminated before yielding any linguistic result: all of the infants died before uttering a single word. The emperor had stumbled upon something remarkable: that "children could not live without clappings of the hands, and gestures, and gladness of countenance, and blandishments." *

Eight hundred years later, in the 1940s, psychoanalyst René Spitz reported on infants caught in a repetition of Frederick's experiment. Spitz described the fate of orphaned children reared in foundling homes and institutions, as well as babies separated from young mothers in prison. In deference to the newly validated germ theory of disease, institutional babies were fed and clothed, and kept warm and clean, but they were not played with, handled, or held. Human contact, it was thought, would risk exposing the children to hazardous infectious organisms.

Spitz found that while the physical needs of the children were met, they inevitably became withdrawn and sickly, and lost weight. A great many died. In a mortal irony, the babies exhibited a vast

---

* Frederick cannot have been pleased with the outcome of the experiment. He was not a man to be trifled with—Saltimbene reports the king once cut off a notary's thumb for the sin of misspelling his name.

vulnerability to the same infections their isolation was meant to guard against. Forty percent of children who contracted measles succumbed to the virus, for example, at a time when the measles mortality rate in the community outside the institution was .5 percent. "The worst offenders," Spitz wrote, "were the best equipped and most hygienic institutions." Death rates at the so-called sterile nurseries near the turn of the century were routinely above 75 percent, and in at least one case, nearly 100 percent. Spitz had rediscovered that a lack of human interaction—handling, cooing, stroking, baby talk, and play—is fatal to infants.

Why should human contact—"gestures and gladness of countenance"—rank with food and water as a physiologic need? The British psychoanalyst John Bowlby picked up this trail in the 1950s. A natural renegade, Bowlby had barely completed his psychoanalytic training before he launched a revolution against the mother church. His creative blend of Freudian metapsychology and Lorenzian ethology produced *attachment theory*, a model that draws parallels between the bonding behavior of humans and animals. Bowlby theorized that human infants are born with a brain system that promotes safety by establishing an instinctive behavioral bond with their mothers. That bond produces distress when a mother is absent, as well as the drive for the two to seek each other out when the child is frightened or in pain. The same behavioral template is manifest in other young mammals, who also cry and cling and seek out their mothers when danger looms.

At the time, Bowlby's ideas were scandalous. The Freudians viewed the mother-infant bond as the "cupboard of love": an infant values his mother because she gratifies his id, as she does when she feeds him. Bowlby's biological bonding system and its infringement on the id's supremacy infuriated the psychoanalysts. They alternately denounced him as naïve and a blasphemer. After Bowlby published his pivotal paper "The Nature of the Child's Tie to His

Mother," Anna Freud rebuked him in frosty and regal tones: "[W]e do not deal with happenings in the external world as such but with their repercussions in the mind." These are fighting words. Accusing a psychoanalyst of realism is verbal annihilation, like calling a composer tone-deaf or a surgeon ham-fisted. Donald Winnicott, English pediatrician turned psychoanalyst and then-president of the British Psychoanalytic Society, wrote that Bowlby's theories were giving him "a kind of revulsion." Even Bowlby's own therapist, Joan Riviere, rose to condemn him at one of the psychoanalytic meetings called for that purpose.

In Bowlby's day, nearly all American psychoanalysts were psychiatrists, and vice versa. While Spitz and Bowlby struggled against the orthodoxies of one profession, their psychologist colleagues in America were burdened by a different but no less restrictive ideology. Psychology, the nonmedical branch of the behavioral sciences, operated for decades under the iron rule of behaviorism. Psychological models of the mother-infant relationship bore the stamp of that withering reign. Reward and punishment, the twin monoliths that taught pigeons to peck levers and rats to run mazes, were invoked as all-purpose tools for shaping human relatedness. Behaviorists advised parents to treat their babies like unruly lab animals. Comforting crying infants was verboten; rewarding distress with attention, they taught, merely reinforces and promotes noxious displays of whining. "Mother love is a dangerous instrument," cautioned the renowned behaviorist John Watson, maintaining that parental affection usually transforms healthy children into contemptible emotional invalids. "Never hug and kiss them," he advised parents, "never let them sit in your lap. If you must, kiss them once on the forehead when they say goodnight."

Harry Harlow's famous work in the fifties dealt synchronous hammer blows to the Freudian and the Pavlovian models of relatedness. In an experiment destined for perpetual notoriety in the pages of college textbooks, Harlow offered young monkeys a choice of two surrogate mothers: a wire mesh cylinder outfitted with a feeding milk bottle, and a terrycloth figure that offered no nutritive sustenance. Without fail, the immature monkeys frequented the wire mother only long enough to dine and treated the furry mother as Mom: they clasped her, squealed at her, embraced her, hid behind her when alarmed. Milk, whether a reinforcing reward or an id-satisfying elixir, failed spectacularly to establish any bond. In trial after trial, the more a doll could be made to resemble a mother monkey, the more infatuated the little monkeys became.

Only Bowlby's attachment theory, which held that proximity to the mother herself is an inborn need, fit the facts. In his view, an infant is born with few motor skills, and so, when his mother strays, he can keep her near by crying—a genetically inherited clarion call that makes a normal mother seek *him* out. As a baby develops muscular coordination, attachment behaviors become more elaborate: a child reaches, grasps, beckons, crawls, or clamors to bring his mother close. Attachment behaviors are clumsy and sputtering in their initial forms, as most behaviors are, but over time they become part of a fluent interaction between child and mother. Children express their separation-related distress first in a nonspecific bleat, and later in pointed communications ("I want you to hold my hand *now.*"). But even crying is not as general a signal as one might suppose: an infant's hunger cry has a unique sound signature. When a mother reaches for the bottle and not the diaper when her baby cries, she is more than guessing about what her child needs.

Certain conditions elicit forceful expressions of a child's instinctive desire to be at a parent's side: unfamiliar places, people, or things; fear, pain, cold, illness, and imposed separations. Adults evidence the same template, although we rarely recognize its outline.

But fear's propensity to amplify bonding is what drives high school couples to see scary movies together. An identical mechanism weaves the ties between people who share a traumatic experience, as in wartime or a disaster. Designers of boot camps, and fraternity and sorority initiations, with varying degrees of consciousness exploit the same process to forge affiliations between dissimilar strangers who must be made to cohere.

Children show fewer outward markers of attachment as they grow up. An eight-year-old is less likely to hold his mother's hand in a department store than a four-year-old, and a fourteen-year-old may not be willing to hold a parent's hand under any circumstances. But the underlying bond endures. An attachment can flourish without overt sign until a disruptive event brings out its expression. People hug each other on departures and arrivals—an act so familiar we might think it nothing more than a custom. But this style of embrace contains silent evidence of attachment: an imposed separation, or the threat of one, reflexively makes people want to reestablish skin-to-skin contact.

The Pliant Years

Psychiatrists are notorious for claiming that pivotal events in the first years of life determine personality. Some skeptics regard this assertion with suspicion, but the study of human attachment has proven it true.

More than twenty years ago, developmental psychologist Mary Ainsworth investigated mothers and their newborn infants and found that the kind of mother a baby has predicts his emotional traits in later life. She first watched how mothers looked after babies and divided caretaking styles into three categories. A year later, Ainsworth then tested the children's emotionality by observing their response to brief separations. A mother who had been consistently attentive, responsive, and tender to her infant raised a *se-*

cure child, who used his mother as a safe haven from which to explore the world. He was upset and fussy when she left him and reassured and joyful when she came back. A cold, resentful, rigid mother produced an *insecure-avoidant* child, who displayed indifference to his mother's departures and often pointedly ignored her on her return, turning his back or crawling away to a suddenly fascinating toy in the corner. The baby of a mother distracted or erratic in her attentions became an *insecure-ambivalent* toddler, clutching at his mother when they were together, dissolving into wails and shrieks when the two were separated, and remaining inconsolable after their reunion.

As the children matured, mothers' parenting aptitude predicted more and more budding personality traits. Babies of responsive mothers developed into grade-schoolers who were happy, socially competent, resilient, persistent, likable, and empathic with others. They had more friends, were relaxed about intimacy, solved problems on their own when they could, and sought help when they needed it. Infants reared by the cold mothers grew up to be distant, difficult-to-reach kids who were hostile to authority, shunned togetherness, and wouldn't ask for comfort, particularly when they were hurt. They often had a mean streak and seemed to take pleasure in provoking and upsetting other children. The offspring of the unpredictable mothers metamorphosed into children who were socially inept, timid, hypersensitive, and lacking in confidence. Hungry for attention and easily frustrated, they frequently asked for assistance with simple tasks that should have been within their competence.

The modest study that Ainsworth began has since swelled into a mountain of meticulous investigation. Long-term data are still rolling in; children have been followed from infancy to their teenage years. Attachment security continues to be a powerful predictor of life success. The securely attached children have a considerable edge in self-esteem and popularity as high school students, while the insecurely attached are proving excessively susceptible to the sad ensnarements of adolescence—delinquency, drugs, pregnancy, AIDS. Almost two decades after birth, a host of academic, social, and personal variables correlate with the kind of mother who gazed down at her child in the cradle.

Ainsworth (and the many researchers who followed her) proved that what a mother does with her baby *matters.* Mothers shape their children in long-lasting and measurable ways, bestowing upon them some of the emotional attributes they will possess and rely on, to their benefit or detriment, for the rest of their lives. The results of this research accord agreeably with common sense. If raising children entails any talent or skill, if one supposes that parenting is more neurally complicated than a reflex, then some people will inevitably possess a greater adeptness at nurturing emotionally healthy children. And from the study of attachment, we can learn who these parents are and how they do what they do.

Ainsworth found no simple correlation between the length of time a mother spent attending to her child and his ultimate emotional health. The securely attached children were not necessarily the infants who were taken up into their mothers' arms most frequently or held the longest. Ainsworth observed instead that secure attachment resulted when a child was hugged when he wanted to be hugged and put down when he wanted to be put down. When he was hungry, his mother knew it and fed him; when he began to tire, his mother felt it and eased his transition into sleep by tucking him into his bassinet. Wherever a mother sensed her baby's inarticulate desires and acted on them, not only was their mutual enjoyment greatest, but the outcome was, years later, a secure child.

By the grace of what miraculous intermediary do mothers know when to approach an infant and when to let him be, when a baby needs the warmth of her embrace and when he needs room to breathe? Limbic resonance gives her the means to that telepathy. By looking into his eyes and becoming attuned to his inner states, a mother can reliably intuit her baby's feelings and needs. The regular application of that knowledge changes a child's emotional makeup. The precise details of that process are now coming to light, as the neural systems underlying relatedness yield up some of their secrets. Bowlby thought the goal of attachment was to establish physical security for an infant, whose helplessness requires a nearby protector. His thinking was audacious in its day, but the reach of relationships is far greater than he imagined. Investigations into the physiology of relatedness now tell us that attachment penetrates to the neural core of what it means to be a human being.

THE ANATOMY OF LOVE

Mourning Becomes Electric

Take a puppy away from his mother, place him alone in a wicker pen, and you will witness the universal mammalian reaction to the rupture of an attachment bond—a reflection of the limbic architecture mammals share. Short separations provoke an acute response known as *protest*, while prolonged separations yield the physiologic state of *despair.*

A lone puppy first enters the protest phase. He paces tirelessly, scanning his surroundings from all vantage points, barking, scratching vainly at the floor. He makes energetic and abortive attempts at scaling the walls of his prison, tumbling into a heap with each failure. He lets out a piteous whine, high-pitched and grating. Every aspect of his behavior broadcasts his distress, the same discomfort that all social mammals show when deprived of those to whom they are attached. Even young rats evidence protest: when their mother is absent they emit nonstop ultrasonic cries, a plaintive chorus inaudible to our dull ape ears.

Human adults exhibit a protest response as much as any other mammal. Anyone who has been jilted in an infatuation (i.e., just about everybody) has experienced the protest phase firsthand—the inescapable inner restlessness, the powerful urge to contact the person ("just to talk"), mistaken glimpses of the lost figure everywhere (a seething combination of overly vigilant scanning and blind hope). All are part of protest. The drive to reestablish contact is sufficiently formidable that people often cannot resist it, even when they understand that the other person doesn't want anything to do with them. Human beings manifest searching and calling in lengthy letters, frantic phone calls, repeated e-mails, and telephoning an answering machine just to hear another's voice. The tormented letter that a rejected lover composes turns out to be an updated version of a baby rat's constant peep: the same song, in a slightly lower pitch.

BEHAVIOR		PHYSIOLOGY
Increased		**Increased**
Motor activity		Heart rate
Vocalization		Body temperature
Demeanor		Catecholamine synthesis
Searching		Cortisol synthesis

The behavior and physiology of the protest phase. (Adapted from Hofer, 1987.)

A mammal in protest shows a distinct physiology. Heart rate and body temperature increase, as do the levels of *catecholamines* and *cortisol*. Catecholamines (like adrenaline) elevate alertness and activity. A young mammal who has lost his mother ought to stay alert long enough to find her, and the rise in catecholamines during protest promotes his vigil. This part of the ancient attachment machinery may also keep a human being staring at the ceiling all night after a breakup. Cortisol is the body's major stress hormone, and its sharp elevation in separated mammals tells us that relationship rupture is a severe bodily strain. Cortisol levels rise sixfold in some mammals after just thirty minutes of isolation.

The Heart's Discontent

A lone puppy's protest phase doesn't last forever. Reunite the pup with his mother, and protest terminates. If the separation is prolonged, a mammal enters the second stage: *despair.* Like protest, despair is a coherent physiologic state—a set of behavioral inclinations and bodily reactions common to mammals. Despair begins with the collapse of fretfulness into lethargy: the animal stops his back-and-forthing, stops whimpering, and curls up in a despondent lump. He drinks little and may show no interest in food at all. If a peer or playmate is introduced into the pen, he may regard him with a bleary eye and turn away. He will have a slumped, dejected-looking posture and a sad facial expression. As the universality of emotional expressiveness lets us know, a mammal in despair looks miserable.

The physiologic signature of the despair phase is that of widespread disruption of bodily rhythms. Heart rate will be low, and on the electrocardiogram we will find abnormal, serrated beats in-

An isolated rhesus monkey. (From *Kaplan and Sadock's Synopsis of Psychiatry,* Eighth Edition. Reprinted with permission of Lippincott, Williams & Wilkins.)

truding into the regular procession of slender spikes that demarcate a healthy heart's metronomic cadence. Sleep will change considerably: lighter, with less dreaming or REM sleep, and more spontaneous nocturnal awakenings. Circadian rhythms, which coordinate the rise and fall of physiologic parameters with the light-dark cycle of the day, will also shift. The level of growth hormone in the blood will plummet. Even immune regulation undergoes major alterations in response to prolonged separation.

Anyone who has grieved a death has known *despair* from the inside: the leaden inertia of the body, the global indifference to everything but the loss, the aversion to food, the urge to closet oneself away, the inability to sleep, the relentless grayness of the world. Grief can give some insight into what it is like to have a major depression. Despair and depression are close cousins, enough so that despair in laboratory animals is often used as a model for human depressive illness. The disease state we call major depression in human beings may be a twisted variant of the despair reaction. But how and why the neural adaptations to loss can be unleashed inside the brain absent the usual trigger, death of a loved one, remains unknown.

Prolonged separation affects more than feelings. A number of somatic parameters go haywire in despair. Because separation deranges the body, losing relationships can cause physical illness. Growth hormone levels plunge in despair—the reason why chil-

BEHAVIOR		PHYSIOLOGY	
Decreased		Decreased	
Motor activity	Vocalization	Heart rate	O₂ Consumption
Socialization	Food/Water intake	Body temp. & wt.	REM Sleep
Play	**Demeanor**	Growth hormone	Cellular Immunity
Increased	Slouched posture	Increased	
Self-huddling	Sad facial expression	Sleep arousals	Irregular heartbeat

The behavior and physiology of the despair phase. (Adapted from Hofer, 1987.)

dren deprived of love stop growing, lose weight no matter what their caloric intake, and dwindle away. Children confined to a hospital for extended periods of time used to surrender to this syndrome in droves. René Spitz called their affliction "hospitalism," a term overtaken by the politely tautological phrase still employed, "failure to thrive." Once doctors appreciated the physical damage contained in social loss, they increased the survival of these children simply by allowing them more contact with their parents.

Children aren't the only ones whose bodies respond to the intricacies of loss: cardiovascular function, hormone levels, and immune processes are disturbed in adults subjected to prolonged separation. And so medical illness or death often follows the end of a marriage or the loss of a spouse. One study, for instance, found that social isolation tripled the death rate following a heart attack. Another found that going to group psychotherapy doubled the postsurgical lifespan of women with breast cancer. A third noted that leukemia patients with strong social supports had two-year survival rates more than twice that of those who lacked them. In his fascinating book *Love & Survival,* Dean Ornish surveyed the medical literature on the relationship between isolation and human mortality. His conclusion: dozens of studies demonstrate that solitary people have a vastly increased rate of premature death from all causes—they are three to five *times* likelier to die early than people with ties to a caring spouse, family, or community.

With results like these backing the medical efficacy of mammalian congregation, you might think that treatments like group therapy after breast cancer would now be standard. Guess again. Affiliation is not a drug or an operation, and that makes it nearly invisible to Western medicine. Our doctors are not uninformed; on the contrary, most have read these studies and grant them a grudging intellectual acceptance. But they don't *believe* in them; they can't bring themselves to base treatment decisions on a rumored phantom like attachment. The prevailing medical paradigm has no capacity to incorporate the concept that a relationship *is* a physiologic process, as real and as potent as any pill or surgical procedure.

THE HIDDEN PERSUADERS

Science is an inherent contradiction—systematic wonder—applied to the natural world. In its mundane form, the methodical instinct prevails and the result, an orderly procession of papers, advances the perimeter of knowledge, step by laborious step. Great scientific minds partake of that daily discipline and can also suspend it, yielding to the sheer love of allowing the mental engine to spin free. And then Einstein imagines himself riding a light beam, Kekule formulates the structure of benzene in a dream, and Fleming's eye travels past the annoying mold on his glassware to the clear ring surrounding it—a lucid halo in a dish otherwise opaque with bacteria—and penicillin is born. Who knows how many scientific revolutions have been missed because their potential inaugurators disregarded the whimsical, the incidental, the inconvenient inside the laboratory?

In 1968, Myron Hofer (now professor of psychiatry and director of the Division of Developmental Psychobiology at Columbia University) was looking into the brain's control over heart rate when felicitous accident struck. He came to work one morning to find that a freedom-loving mother rat had chewed through her cage and escaped during the night. Hofer happened to notice that her abandoned litter of pups showed heart rates less than half normal. He surmised that the pups' cardiac cells had cooled without a mother's warmth, and decided to run a test on his idle hypothesis. He provided lone baby rats with a heat source that mimicked a maternal presence. To his surprise, the hearts in the pups beat just as slowly before warming as after. Somehow, a mother rat possesses an organic thermoregulatory power that disembodied heat does not.

Intrigued by this mysterious maternal force, Hofer set about divining the arcane physiology of orphaned rat pups. In experiment after experiment, he replaced the missing mother with single fragments of her sensory qualities. A piece of cloth with her scent on it, a lamp that radiated heat at her body temperature, strokes on a pup's back with a brush, simulating her grooming—Hofer used them as deliberately partial substitutes for a mother rat.

Hofer found that restoring a *single* maternal attribute could prevent just *one* physiologic aspect of despair, without affecting any of the others. A mother's body warmth and olfactory cues direct her infant's activity level, while her tactile stimulation determines her pup's growth hormone level. Milk delivery to a pup's stomach fixes its heart rate, while the periodicity of feeding modulates sleep-wake states.

Hofer realized not only that the tie linking a mother rat to her baby is vital and corporeal, but also that the bond itself is woven from separate strands, each a distinct regulatory pathway in the body. A mother continuously adjusts her infant's physiology. One can interrupt a single thread of her influence and disrupt the corresponding physiologic parameter in her baby. When the mother is

INFANT SYSTEM		Decreased by	Increased by	
Behavioral	Activity level	body warmth	tactile, olfactory cues	
Sucking	Nutritive	nutrient (distention) tactile (perioral)	signal unknown	
Neurochemical	Norepinephrine/ Dopamine levels	tactile & olfactory	body warmth	M A T E R N A L R E G U L A T O R
	ODC levels	sensorimotor	tactile	
	Opiate levels		sensorimotor	
Metabolic	O₂ consumption	signal unknown	sugar content of milk	
Sleep	REM		milk periodicity; tactile	
	Arousals	milk periodicity; tactile		
Cardiovascular	Heart rate		milk	
	Vasoconstriction	milk		
Endocrine	Growth hormone		tactile, body warmth	
	Corticosterone	tactile, milk		
Immune	B-cell and T-cell response	signal unknown	signal unknown	
Circadian	Phase set		milk, body warmth	
	Period length	melatonin?	melatonin?	

Hidden regulators in a rat relationship. (Adapted from Hofer, 1987.)

absent, an infant loses all his organizing channels at once. Like a marionette with its strings cut, his physiology collapses into the huddled heap of despair.

The figure below demonstrates the disharmony that motherlessness unleashes in the bodily rhythms of a baby rat.

Once separated from their attachment figures, mammals spiral down into a somatic disarray that can be measured from the outside and painfully felt on the inside. The rate of disintegration differs—infants are most dependent upon external support, and without it they lapse quickly. The stability of older children decays more slowly, and that of many adults, slower still. Whatever the age, the eventual slide is inevitable; the physiology of social mammals is unstable at any speed. Hofer's delineation of this frailty opened the door to a novel view of human relatedness.

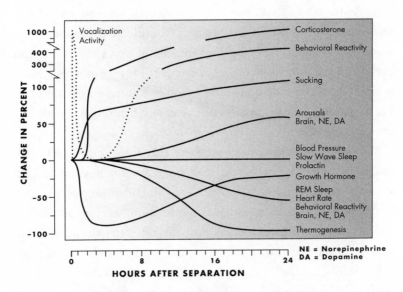

Physiologic chaos unleashed by separation. (From *Attachment Theory: Social, Developmental, and Clinical Perspectives,* edited by S. Goldberg, R. Muir, and J. Kerr, 1995. Reprinted with permission of the Analytic Press.)

THE OPEN CIRCLE

Most people assume that the body they inhabit is *self*-regulating—that their own physiologic balance occurs within a *closed loop*. Cruise control is the classic example of a closed loop, with a car's self-contained system checking its speed and adjusting its own throttle accordingly. An auto under manual operation, on the other hand, is one half of an *open-loop* duo—here the car rolls on; an utterly external agent takes in the speed of the rushing landscape and, with his feet pushing pedals, instigates throttle changes; the car's velocity rises and falls. A car minus cruise control is no master of its fate; alone, it cannot maintain any desired speed above zero.

Is the human body an open- or closed-loop affair? Do we possess internal cruise control analogues that monitor and modify our physiologic oscillations, or is someone else in the somatic driver's seat? Both, to an extent. Some of our somatic systems are closed, self-regulating loops. Others are not. Consider, for instance, that women who spend time together frequently find their menstrual cycles coming into spontaneous alignment. This harmonious, hormonal communion demonstrates a bodily connection that is limbic in nature, because close friends achieve synchrony more readily than those who merely room together.

A number of scientists now believe that somatic concordances like these are not just normal but necessary for mammals. The mammalian nervous system depends for its neurophysiologic stability on a system of interactive coordination, wherein steadiness comes from synchronization with nearby attachment figures. Protest is the alarm that follows a breach in these life-sustaining adjustments. If the interruption continues, physiologic rhythms decline into the painful unruliness of despair. Evolution has given mammals a shimmering conduit, and they use it to tinker with one another's physiology, to adjust and fortify one another's fragile neural rhythms in the collaborative dance of love.

We call this mutually synchronizing exchange *limbic regu* human body constantly fine-tunes many thousands of phy parameters—heart rate and blood pressure, body temperature mune function, oxygen saturation, levels of sugars, hormones, sa ions, metabolites. In a closed-loop design, each body would self monitor levels and self-administer correctives, keeping its solitary system in continuous harmonious balance.

But because human physiology is (at least in part) an open-loop arrangement, an individual does not direct all of his own functions. A second person transmits regulatory information that can alter hormone levels, cardiovascular function, sleep rhythms, immune function, and more—inside the body of the first. The reciprocal process occurs simultaneously: the first person regulates the physiology of the second, even as he himself is regulated. Neither is a functioning whole on his own; each has open loops that only somebody else can complete. Together they create a stable, properly balanced pair of organisms. And the two trade their complementary data through the open channel their limbic connection provides.

A baby's physiology is maximally open-loop: without limbic regulation, his vital rhythms collapse, and he will die—as Frederick II and René Spitz both proved. In current parlance, babies outsource most physiologic governance to parents and gradually bring those duties back in-house over months to years. Their early exposure to the external order that parents provide teaches babies how to manage some physiologic rhythms on their own. Two studies, for instance, compared premature infants who slept with a standard teddy bear to those supplied with a "breathing" bear—an ordinary stuffed animal connected to a ventilator and set to inflate and deflate at a rhythmic fraction of the baby's own respiratory rate. The infants with the breathing bear later showed more quiet sleep and more regular respiration than those who slept with a static Winnie-the-Pooh. Regular sighs taught the preemies respira-

ʌdern technology providing the means to an
ʌation.

matures, a baby reclaims some regulatory
ʌem autonomously. Even after a peak par-
never transition to a fully self-tuning
ʌemain social animals: they continue to require
ʌr stabilization outside themselves. That open-loop de-
ʌıgn means that in some important ways, people cannot be stable
on their own—not should or shouldn't be, but *can't* be. This
prospect is disconcerting to many, especially in a society that prizes
individuality as ours does. Total self-sufficiency turns out to be a
daydream whose bubble is burst by the sharp edge of the limbic
brain. Stability means finding people who regulate you well and
staying near them.

Taking a rhesus monkey away from his mother too soon or subjecting him to lengthy maternal absences will produce a monkey
with a lifelong vulnerability to despair. Limbic regulation explains
why: with less internalized capacity for self-supervision, such a
mammal slips precipitously into physiologic chaos whenever his
extrinsic source of stability moves out of range. Human children
of erratic mothers are clingy for the same reason. Because they
haven't been able to absorb sufficient closed-loop control over their
physiology, they need to stay near an external regulator to remain
in balance.

This necessary intermingling of physiologies makes relatedness
and communal living the center of human life. We recognize in-
stinctively that healthy humans are not loners. Of his famous retreat to Walden Pond, Thoreau wrote, "I went to the woods
because I wished to live deliberately, to front only the essential
facts of life," but he did not front them alone. His nearest neigh-
bor was but a mile distant, and Concord two miles; Thoreau de-
pended on both liberally, and dined frequently with friends. In

children's stories and in life, disease creates hermits and cabin-dwelling Kaczynskis. Limbic regulation makes expulsion from the company of others the cruelest punishment human beings can devise. When his friend Friar Laurence tells Romeo that his death sentence has been commuted to interminable exile, Romeo's heart prepares to break:

> *And say'st thou yet that exile is not death?*
> *Hadst thou no poison mix'd, no sharp-ground knife,*
> *No sudden mean of death, though ne'er so mean,*
> *But "banished" to kill me? "Banished"?*
> *O friar, the damned use that word in hell;*
> *Howlings attend it: how hast thou the heart,*
> *Being a divine, a ghostly confessor,*
> *A sin-absolver, and my friend profess'd,*
> *To mangle me with that word "banished"?*

The Outsiders

Limbic regulation mandates interdependence for social mammals of all ages. But young mammals are in special need of its guidance: their neural systems are not only immature but also growing and changing. One of the physiologic processes that limbic regulation directs, in other words, is the development of the brain itself—and that means attachment determines the ultimate nature of a child's mind. The importance of limbic contact for normal brain development shows itself most starkly in the devastating consequences of its omission.

Feed and clothe a human infant but deprive him of emotional contact and he will die. But infant monkeys are hardier than humans in the face of such privation. Monkeys reared without their mothers often survive, but their neural systems are permanently maimed.

Gary Kraemer, professor and chair of the Department of Kinesiology at the University of Wisconsin, and a leading researcher on the neurobiology of social deprivation, has described and investigated the effects of what is termed the *isolation syndrome.* Monkeys raised alone cannot engage in reciprocal interactions with normal monkeys, who consistently reject them. They are unable to mate. If solo-reared females undergo artificial impregnation, they show a striking lack of mammalian attitudes toward their infants: indifference and neglect alternate with savage attacks. Isolates are unpredictably vicious to adults as well. Ordinary monkeys usually break off a conflict when dominance has been settled, but those reared in isolation often fight to the death and beyond, rending and dismembering opponents. Self-mutilation is another of solitude's legacies: these monkeys bite their own arms, bang their heads against the wall, and gouge out their eyes. Social environment even fixes the normal formation of such behavioral basics as eating and drinking: isolates typically engage in prolonged food and water binges.

An isolated monkey becomes a grotesque caricature because the mammalian nervous system cannot self-assemble. Many subsystems of the mammalian brain do not come preprogrammed; maturing mammals need limbic regulation to give coherence to neurodevelopment. Without this external guidance, neural cacophony ensues: behavioral systems are constructed, but without proper harmony between the interdigitating parts. Like the isolates described above, mammals that grow up in the absence of central coordination are jagged and incomplete. Their brains produce fractured behaviors that emerge at the wrong times, in the wrong places, in the wrong ways. They have aggression, for instance, but not the modulated, momentary fierceness that serves to challenge or defend a place in the pecking order. Instead they show wild swings of unpredictable violence incompatible with membership in a social group. A monkey cannot even grow up knowing how to eat or drink in a balanced way unless his mother was at his side during childhood.

Love, and the lack of it, change the young brain forever. The nervous system was once thought to unfold into maturity in accordance with the instructions in its DNA, much as a person alone in a room might, with a set of directions and a flurry of creases, produce an origami swan. But as we now know, most of the nervous system (including the limbic brain) needs exposure to crucial experiences to drive its healthy growth. In work that netted the 1981 Nobel Prize for medicine, David Hubel and Torsten Wiesel showed that kittens raised with one eye covered grew up to be cats with marked aberrancies in the brain areas serving vision. The same holds true for the neural systems that direct limbic resonance and regulation: relevant experience is a necessary part of the process that leads to the brain's final structure. The lack of an attuned mother is a nonevent for a reptile and a shattering injury to the complex and fragile limbic brain of a mammal.

Raising a monkey in seclusion provides direct data on the neural effects of *total* social deprivation. Human infants almost never survive such drastic conditions. To evaluate the impact of subtler bonding derangements, a group of researchers devised an ingenious way to make healthy monkeys into poor mothers. They place monkey mother and infant in an environment where food is not always readily available. Sometimes the mother has easy access to nutrients; at other times she must search diligently to find enough to feed herself and her baby. The unpredictability of that circumstance preys on the mother's mind and erodes her parental attentiveness.

Such distracted, apprehensive mothering endows juvenile monkeys with emotional vulnerabilities and altered neurochemistries. The monkeys so raised show magnified despair and anxiety reactions, and their brains reveal changes in the neurotransmitter systems that control these emotional states. Unlike the sweeping harm that isolation rearing unleashes, these defects are focal and faint enough that a mother's presence can mask them: at her side, the impaired young monkey seems normal. But separate the two and his apparent stability evaporates—a condition called *pseudoindependence.*

Full-grown, these monkeys are living proof of limbic regulation's enduring power: they are timid, clingy, subordinate, and clumsy in their efforts to establish ties to other monkeys. The brains of these animals evidence permanent alterations in neurochemistry. Just because their mothers once lived under a pall of uncertainty, these adult animals show lifelong changes in levels of neurotransmitters like serotonin and dopamine. With their vulnerability to anxiety and depression, their social awkwardness and failures to attach as adults, these monkeys exhibit a close animal counterpart to the multifaceted misery that in human beings is labeled neurotic.

Despite the centrality of limbic regulation, not all mammals live to link and link to live. Giant pandas spend their lumbering and bamboo-munching days alone, and come together only for the essential sexual union that preserves their species. Even the order of the great apes has a member that is at best semisocial: the orangutan, whose male members find one another so intolerable they cannot manage a peaceful assembly. Only a mother orang and her offspring can stand each other for any appreciable period of time.

How are we to make sense of these apparently casual desertions of the organizing principle of mammalian life? The meandering path of evolution supplies an answer. When necessity hatches a breed of organisms with a novel skill, some of them may, in time, find it advantageous to abandon their hard-won heritage and resume their former lives. Thus the world contains reptiles that have returned to the same sea from which their fishy forebears labored to escape, and birds that relinquished the skies so long ago their wings have shrunk to aeronautically useless flaps. Among this group of backward creatures are the asocial mammals: furry, milk-bearing beings whose ancestors huddled together as families, and who have slipped back into a solitary way of life older still.

BUILDING BLOCKS OF LOVE

When people have trouble with their emotions—a bout of anxiety or depression, say, or seasonal gloominess—they often want science to pinpoint an offending neurotransmitter in the way that a witness picks the perp out of a lineup. Is it excessive norepinephrine, too little dopamine, errant estrogen? The answer is apt to dissatisfy: no single suspect can be fingered with confidence because the question itself attributes a fallacious simplicity to the brain.

When trying to fathom an immense, intricate system, drawing direct arrows of causality between micro- and macrocomponents is perilous. Which stock caused the crash of '29? Which person triggered the outbreak of World War I? Which word of Poe's "The Raven" suffuses it with an atmosphere of brooding melancholy? Neuroscientists understand the immediate chemical effect of a handful of medications, but connecting the dots of those minute molecules to sketch human actions, thoughts, feelings, and traits means tracing a baffling, blossoming tangle of biochemical events. The brain's dense thicket of interrelationships, like those of history or art, does not yield to the reductivist's bright blade.

Statements reading "Chemical A causes Human Trait B" have no meaning, despite their popular appeal. The brain is no simple machine, with a lever here releasing joy and a pulley there prompting panic. We can nonetheless wring valuable information about relatedness from neurochemistry. Neurotransmitters are not created equal, and some are far more important than others in directing limbic functions, including love. Ongoing investigations have implicated three crucial chemical players: serotonin, opiates, and oxytocin.

THE NOTORIOUS TRANSMITTER

Medical science chanced upon antidepressants in the 1950s, and for thirty years most physicians were too frightened to prescribe them in sufficient quantities to permit their efficacy. The reason was simple: conventional antidepressants were among the easiest drugs one could use to kill oneself. In many cases, a meager week's worth of medication was lethal enough to effect a suicide. In 1988, when Eli Lilly introduced an antidepressant that didn't kill people even when taken in bulk, relieved physicians began prescribing it like crazy. Within months, Lilly's medication became the most widely prescribed antidepressant in the world—the infamous Prozac, the drug that made *serotonin* a household word.

Originally conceived as a treatment for depression, Prozac and the other serotonin agents soon proved to be multifunctional molecules with a host of unforeseen and beneficial uses. As tens of millions of patients tried these medications, the adventitious effects piled up. Anxiety, hostility, stage fright, PMS, road rage, bulimia, low self-confidence, premature ejaculation—and the proclivity of restless dogs to lick the hair off their forelimbs—all are potentially remediable through some judicious tinkering with a few of the brain's many serotonin circuits. A lesser-known property of the serotonin agents is that they sometimes attenuate the pain of loss. It doesn't happen to everyone, but a select group derives benefit from serotonin agents because they weaken the heartache that comes from losing someone.

One person we know, for instance, was trapped in a dismal relationship simply because she could not get around the pain of loss. No matter how much unhappiness her mate caused her, at every attempt to break with him a taller wave of wretchedness welled up inside her. And so her inner scales regularly tipped in favor of staying with a man who could not satisfy her. "I want to stop *so badly* with him," she said. "Our relationship goes on and on, and I keep thinking, 'This time, it's over,' but it's never going to be. I feel like moving across the country just to get away from him because it's been going on for so long—I'm fighting with myself constantly over it. I tell myself, 'Just go away, don't ever contact him again,' and I can't. I *can't.*" Years of therapy clarified her misery but did not diminish it. But when she took a serotonin agent, the balance of her sorrows shifted slightly. Loss hurt a little less. She did then what she had been unable to do: leave her lover without intolerable suffering.

The freedom to leave a relationship is a bequest, not a birthright. As the burgeoning research on primate attachments tells us, early nurturance can stretch forward in time to insulate adults from the destabilizing pangs that solitude brings. As a society, if we do not attend to the limbic needs of our own young, we risk creating an epidemic of loss vulnerability. Serotonin agents will then become not just a remedy to retrieve those few teetering on the brink of desolation's abyss, but a way of life for a culture that has settled on the lip of the precipice itself.

THE RELIGION OF THE MASSES

The juices of the flowering plant *Papaver somniferum* possess an exceptional quality: they alleviate pain. Scrape and dry the poppy's exudations and the result is opium—a mixture of homologous compounds from the opiate dynasty, an extended chemical family that includes such notables as morphine, heroin, and laudanum. *Papaver's* extract eliminates pain because the selfsame opiates are vital components of the brain's own analgesic system. Prompt deliverance from physical torment was miraculous to the first physicians who dispensed it. Thomas Sydenham said in 1680: "Among the remedies which it has pleased Almighty God to give to man to relieve his sufferings, none is so universal and so efficacious as opium."

Sydenham was telling only half the story. Opiates not only extinguish the pain that comes from physical wounds but they also erase the emotional excrucitation arising from the severing of a relationship. The limbic brain has more opiate receptors than any other brain area, perhaps for this purpose. Separation studies attest to the brisk effectiveness of opiates as anesthetics of loss: if a dam is taken away from her puppies, their distress erupts. Give them a tiny dose of opiate (too small to be sedating), and the pups' protest vanishes.

Poets and other disreputable types have known about this power for thousands of years. The fourth book of Homer's *Odyssey* contains this medically accurate description of a dinner party where the conversation has taken a sorrowful turn to talk of lost comrades:

> *A twinging ache of grief rose up in everyone . . .*
> *But now it entered Helen's mind*
> *to drop into the wine that they were drinking*
> *an anodyne, mild magic of forgetfulness.*
> *Whoever drank this mixture in the wine bowl*
> *would be incapable of tears that day—*
> *though he should lose mother and father both,*
> *or see, with his own eyes, a son or brother*
> *mauled by weapons of bronze at his own gate.*

The amelioration of mourning fell to the opiates through the happenstance of biological history. Bodily damage risks death—a stark fact that drove the evolutionary development of a neural system that senses injury. The business end of that brain function is *hurt*—a potent incentive for animals to get out of harm's way. But inside the body's endless opposing rhythms, every physiological tendency exists alongside its polar opposite. And so the brain contains not only neurotransmitters that produce pain but also those that assuage: the opiates. By the time the limbic brain arose, and mammals came to depend on mutual regulation for survival, a refined mechanism was already in place to manage the mental after-effects of physical trauma. Evolution then recruited parts of that system to process the emotional pain of loss.

While a neocortical brain post-Descartes can wax eloquent on the division between mind and body, the other brains draw no such distinction. Damage to one's arm or to one's neurophysiology are equally real and, to a mammal, the latter may be more crippling. What matters most to Pain Central is not the philosophical category a slight belongs to but the level of jeopardy it threatens. Given the open-loop physiology of mammals and their dependence on limbic regulation, attachment interruptions are dangerous. They ought to be highly aversive. And so they are: like a shattered knee or a scratched cornea, relationship ruptures deliver agony. Most people say that no pain is greater than losing someone they love.

The intertwining of loss with the opiates permits the brain to be hot-wired in circumstances of dire need. Psychiatrists often see people who deliberately injure themselves in minor but stinging ways—like making shallow razor cuts to the forearm or cigarette burns to the thigh. These individuals have garnered a multiplicity of polysyllabic labels over the years, and their self-destructive bent has been ascribed to various convoluted motives: a desire for attention, an attempt to manipulate, a turning of anger against the self.

Most of them have one thing in common: an exquisite, lifelong sensitivity to separation's pain. The miniature losses contained in a rebuke, a spat, and other transient relationship rifts can arouse in them an unbearable blend of despondency and grief. Then follows an episode of self-harm—a prick, a burn, an incision into the skin. Beneath and within the abused epidermis, palpitating pain fibers send their drumbeat signal to the brain, warning of damage. These messages release pain's counterweight: the blessed, calming flow of opiates, and thus, surcease of sorrow. Chronic self-mutilators provoke the lesser pain to trick their nervous systems into numbing the unendurable one.

Less drastic routes abound: warm human contact also generates internal opiate release. Our lovers, spouses, children, parents, and friends are our daily anodynes, delivering the magic of forgetfulness from the twinging ache of mammalian loneliness. Potent magic indeed.

A Prairie Dog's Life

The third neurotransmitter directing attachment coordinates physiologic events around childbirth—it stimulates uterine contractions and milk ejection—but until recently no one suspected its striking emotional power.

The passionate properties of *oxytocin* have been elucidated inside the brain of an unlikely candidate for scientific fame: the prairie dog. Thomas Insel, psychiatric researcher and the director of the Center for Behavioral Neuroscience at Emory University, has studied two species of prairie dogs (known also as *voles*). Prairie voles (*Microtus ochrogaster*) affiliate: adults are monogamous, both parents nurture their young, and husband and wife spend most of their time sitting side by side. Mountain-dwelling montane voles (*Microtus montanus*) are not nearly so social: mating patterns among these more casually attached rodents tend toward the promiscuous, and parents engage in much less caretaking of the young than do their flatland cousins. Paternal montane voles frequently ignore their offspring, and mothers often abandon their young ones two weeks postpartum.

Insel compared the brains of the two species and noticed they differ in the activity of just one neurotransmitter system—

oxytocin. The limbic brains of the affiliative prairie voles are loaded with oxytocin receptors, while the more aloof montane voles have far fewer. Oxytocin activity rises in the montane vole only around the time of birthing pups, when affiliation is a necessity. After rearing is done for the montane vole, oxytocin falls again, and so does bonding. Mother and her young then go their own ways.

The love lives of prairie dogs implicate oxytocin in forging the bonds of relatedness. Oxytocin levels surge in human mothers around birth—to stimulate labor and nursing, it was thought, but science sees those hormone levels in a new light. Experts have debated for decades about whether mothers and infants form a bond in the hours after parturition, and about the wisdom of separating the two at this time, as was the custom in Western in-hospital childbirth. High oxytocin levels around birth point to a crucial relationship event. They tell us that a mother and her child are meant to be together postpartum, when their neurochemistries are busy weaving the ties between them.

Oxytocin also gushes at puberty, when teenage crushes first bloom. It may seem strange that a simple molecule could initiate infatuation's sweet adolescent spell, but everything that happens in the brain begins with neurochemistry—including the wonder of puppy love, whose complex secret resides within the limbic brains of prairie dogs as surely as it does within ours.

THE BREADTH OF THE BOND

Human beings can decipher some of the limbic manifestations of other mammals, and vice versa. Some emotional communications are species specific: when a cat blinks her almond eyes and looks away, this signal, seemingly so rich to other felines within range, rests safely beyond human ken. But despite the variety of emotional expressiveness among mammals, they partake of a common neural infrastructure. A consequence of this shared limbic inheritance is often taken for granted: different species can attach to one another.

On the average Sunday here in Marin County, ample evidence awaits in front of the nearest grocery store, where one or two golden retrievers will often be tied outside while their owners shop. Most of the time, the dogs are standing up, peering through the glass door, trying to catch a glimpse of the one person inside who means something. From time to time, someone will come by, in or out, and pat a dog on the head. The dog accepts this affection, if a little impatiently. But as his owner heads out the door, he trembles and leaps with unmistakable eagerness. Separation, vigilant scanning, indifference to those outside the bond, reunion, and joy have taken place in a ten-minute span in front of the local market, and all between two species separated in evolutionary time by tens of millions of years.

Somehow the attachment architecture is general enough that a human being and a dog can both fit within the realm of what each considers a valid partner. And the two can engage in limbic regulation: they spend time near each other and miss each other; they will read some of each other's emotional cues; each will find the presence of the other soothing and comforting; each will tune and regulate the physiology of the other. Limbic regulation is life-sustaining. This is why pets can make people not only feel better but also live longer. Several studies have shown dog-owning cardiac patients die at one quarter to one sixth the rate of those who forgo canine companionship.

More than twenty-five years ago, Lewis Thomas wrote, "Although we are by all odds the most social of all social animals—more interdependent, more attached to each other, more inseparable in our behaviors than bees—we do not often feel our conjoined intelligence." The science of our day is allowing us to understand what interdependence is for, to know the intended outcome of the inseparability, to divine the nature of our conjoined state.

We are attached to keep our brains on track, in a process that begins before birth and sustains life until its end. The earliest portion of that duet must catch our attention: attachment changes a young mammal forever, as limbic regulation carves enduring patterns of knowledge into the developing circuits of the mind. To understand how attachment sculpts a person, we need to apprehend *memory*—the process whereby the brain undergoes structural change from experience. Memory does not travel a straight line, and neither does the human heart.

Five

GRAVITY'S INCARNATION

How memory stores and shapes love

Memory is a small word that contains entire worlds. With a minimal exertion of will, anyone can conjure up a vision of places and people long since destroyed by the passage of time, whose impressions remain encoded along winding synaptic paths. Somehow the immensity of the past is dormant within, and parts waken at our command. But memory is more: it defines, creates, and holds a person's mental world together. As a pioneering physiologist of the nervous system, Ewald Hering, saw it:

> *Memory collects the countless phenomena of our existence into a single whole; and as our bodies would be scattered into the dust of their component atoms if they were not held together by the attraction of matter, so our consciousness would be broken up into as many fragments as we had lived seconds but for the binding and unifying force of memory.*

Hering's declaration was prophetic. Every individual lives as a spectral vapor in the neural machine, his thoughts, dreams, feelings, and ambitions the evanescent outcome of intricate signals flowing among billions of neurons. The stability of an individual mind— what we know as *identity*—exists only because some neural pathways endure. The plasticity of the mind, its capacity to adapt and learn, is possible only because neuronal connections can change. The physiology of memory determines the fate of those malleable nodes. It lies at the heart of who we are and who we can become.

A scientific theory of memory is therefore a map of the soul.

Every such diagram must attempt to delineate the mind's Dark Continent: why do people possess emotional knowledge that leaves no conscious trace?

Since time's beginning, romantic partners have searched for each other with exquisite but obscure deliberation. "In literature, as in love," wrote André Maurois, "we are astonished at what is chosen by others." And they are every bit as amazed at us. The very concept of "compatibility" discloses that no all-purpose template for loving predominates. Sexual attractiveness contributes only a minor filter to this selectivity. The number of couples who marry is a minuscule fraction of the many who find each other physically interesting. Not just anyone will do; in fact, to any one person looking for a mate, almost nobody will.

A lover tests the combination of himself plus serial others like a child juxtaposing jigsaw pieces until a pair snaps home. Love's puzzle work is done in the dark: prospective partners hunt blindly; they cannot describe the person they seek. Most do not even realize, as they grope for the geographical outline of a potential piece, that their own heart is a similar marvel of specificity. How do these delicately shaped desires develop? By what means do people learn the discriminating taste that tells them how and whom to love? And why does that knowledge remain opaque to their mind's eye?

Seventy-five years before the memory science elucidated in this chapter, Sigmund Freud proposed a model of unconscious emotional memory that became an institution. Freud's unconscious was a psychic Pandora's box—a repository of thoughts, memories, ideas, and impulses so unpleasant and anxiety-provoking they had to be deleted from consciousness and confined to a mental basement. In concocting his mind cellar, Freud assumed that memories possess the archaeological solidity of a Grecian urn: they can be buried by a sandstorm of repression and, if censorship later weakens, exhumed in pristine condition. "From the repressed memory traces, it can be verified that they suffer no changes even in the longest periods," Freud wrote. "The unconscious, at all events, knows no time limit."

The Pandoran metaphor is enticing. Its central image is pleasingly congruous with the conception of world order handed down from antiquity: sweet reason in the heavens above, malefic monsters below, and the planet's scarred surface a stage for climactic battles between mighty opposites. In practice, the Freudian scheme functions as a bulletproof shield protecting allegations that the unconscious mind contains this or that Boschian beast. If no such creature is ever sighted, one can always attribute the evidential vacancy to repression's strong chains rather than to overactive imaginations. Freud's memory model has thus encouraged many a hair-raising ghost story and no small amount of mischief.

The Franklin case is one such nightmare tale. In the most notorious of repressed memory cases, George Franklin stood trial in 1990 for murder because his daughter Eileen suddenly "remembered" that she saw him beat an eight-year-old girl to death twenty years before. No corroborating witnesses came forward. No physical evidence linked him to the crime—not a fingerprint, fiber, or DNA strand. The lifelike details composing Ms. Franklin's recollected relic had all been published in newspaper accounts decades before. But when a solemn psychiatric expert intoned that Eileen's forgotten "memory" was inarguably bona fide, the jury believed. George Franklin went to prison. After a federal court threw out the conviction five years later, the district attorney quietly elected not to retry the case. His star witness for the prosecution had meanwhile mutilated her tenuous credibility, after "remembering" that her father murdered two other people—crimes that DNA evidence and an airtight alibi proved he could not have committed.

Sunny San Mateo, where the Franklin affair unfolded, is thousands of miles and many decades distant from turn-of-the-century Vienna, where the memory doctrine that was to condemn Franklin originated. Freud himself wielded repression, not to charge people with homicide but to declare the subterranean influence of hidden incestuous thoughts. A child's amorous interest in his parents fixes his attraction to later loves; since revulsion expels this lust from consciousness, he will always be ignorant of his patterned, odious longing. That is Freud's account of unconscious emotional memory. A dramatic yarn, but at least two flaws undermine the model of memory at its center.

First, a memory is not a *thing*. Cardiac muscle fibers are objects, but the heartbeat they generate is a physiologic *event*, a collective flutter that propels life but nevertheless has no mass and occupies no space. A memory is another bodily process, produced by physical objects but itself as immaterial as the soul. If a heart beats once and then rests for a minute, the heartbeat has neither gone someplace nor must it be fetched back against resistance. Memories are the heartbeats of the nervous system, although decades may elapse before any one recurs. They are not objects; they do not travel. And second, modern science has erased Freud's conviction about memory's immutability. Memory is not only mutable, but as we will see in the next chapter, the nature of the brain's storage mechanism dictates that memories *must* change over time.

Because Freud built his model of memory on counterfeit cornerstones, our century has seen it topple. The doctrine of repressed memory has sunk into disrepute; in many courts, "recovered" accusations are no longer admissible evidence of anything. But have no doubt—unconscious emotional knowledge *does* exist. A shadow does lie across the landscape of memory, but that darkness is not the sinister specter of censorship.

When the moon passes directly before the sun, the momentary superimposition throws a circle of twilight on Earth's surface, an *umbra*. When the two resolve again into separate heavenly spheres, the land beholds a second dawn. Freud did not know that the memory voids he was charting are the penumbral shadows of a perpetual mental eclipse. He could not anticipate that someday science would divide memory into two distinct orbs, a sun and moon whose apparent unity misleads. One of the brain's memory mechanisms bathes consciousness in a floodlight of facts and specificity, while the other—older, deeper, quieter—illumines our lives with a pale fire all its own.

THE IMPOSSIBLE DIARY

Who got drunk at your wedding? What color were your first lover's eyes? Which actor starred opposite Myrna Loy in *The Thin Man*? If these questions generate answers, they do so only by the grace of the explicit memory system. Explicit memory, the more public of the brain's twin storage machines, encodes event memories, including autobiographical recollections and discrete facts. When you need access to something you once knew or experienced, a moment's mental sift presents a solution to consciousness. While explicit memory is swift and capacious, a fallacious sense of accuracy attends its frequently erroneous returns. New scanning technologies show that perception activates the same brain areas as imagination. Perhaps for this reason, the brain cannot reliably distinguish between recorded experience and internal fantasy. Oscar Wilde's Miss Prism says in *The Importance of Being Earnest*, "Memory, my dear Cecily, is the diary that we all carry about with us." Sharp Cecily replies, "Yes, but it usually chronicles the things that have never happened, and couldn't possibly have happened."

The hardware that creates explicit memory lies within the brain's temporal area. The most important component is the hippocampus, a graceful spiral of neurons that begins close to the midline and curls its way out into the poles of the temporal lobes.

Nestled deep in the heart of the brain, the hippocampus looks safe from injury. Not so—accidents, strokes, viruses, and neuro-

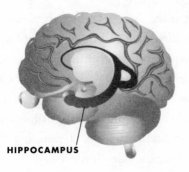

HIPPOCAMPUS

surgical enthusiasm can and do decimate the hippocampus. Patients who have lost their hippocampi bear witness to its memory powers, because no explicit memories can be created without one. These patients suffer from that staple of soap opera story lines, amnesia. Daytime dramas focus on a character's inability to reconstruct an admittedly dizzying array of romantic misadventures, but the real problem for patients without hippocampi is that they cannot record and remember. Their lives are marooned on the island of the present.

One such individual, for instance, we'll call Mr. Underwood: a sixty-seven-year-old man whose family brought him to the hospital because he seemed confused. Subsequent investigation revealed that he had Korsakoff's syndrome—the destruction of some vital pieces of the explicit memory system caused, in this case, by decades of heavy drinking.

With no capacity to recall anything he had seen or done after the damage to his brain, Mr. Underwood became locked into an unchanging present. He always thought it was 1985, and he always thought Ronald Reagan was president. He was continually perplexed to find himself in the hospital, because patiently provided explanations evaporated from his mind within minutes of delivery. His regular doctors and nurses were strangers to him, and every encounter necessitated a fresh round of introductions. He enjoyed telling jokes—often the same joke three or four times in ten minutes, the old man blissfully unaware that repetition had robbed his punchlines of any possibility of punch. He was apt to wander off the ward and roam the hospital in amiable bewilderment until someone took the trouble to examine his wristband and find out who he was and where he came from. That he could never know.

Mr. Underwood had suffered a devastating neurologic injury that confined him to the windowless prison of *now.* A healthy mind flees backward in time every second (*Who is that? Where did I leave the car keys? What did I argue with my wife about last night?*). Doing so enables us to know what we have been through, where we are, what is going on in the world, and why. Information ran through Mr. Underwood like sand through a sieve. He lived a paper-thin existence, skimming by on the surface of time, further and further from knowledge that could ground him in a relevant world.

If Mr. Underwood's drinking had devastated his explicit memory machine, how can he remember anything? How can he remember that Reagan *was* president in 1985, or his wife's name, or his? The hippocampus is a key player in creating explicit memories, but the memories themselves reside elsewhere. Patients like Mr. Underwood can recollect events that happened up until their brain damage (within a few days), and nothing since.

Or so it was thought for many years. Under scrutiny, patients like Mr. Underwood showed researchers that even without explicit memory, the capacity to learn survives. This discovery was like finding a city on the dark side of the moon. The hunt for a hidden second memory system in the brain was on.

COVERT OPERATIONS

A patient like Mr. Underwood was taught to braid—a skill he was unacquainted with before his explicit memory expired. After he had mastered it, experimenters asked him if he knew how to braid. He replied "No," a truthful statement from his point of view. Yet when three strips of cloth were placed in his hands he wove them together without hesitation.

If people form memories without realizing it, how could we ever know? Only by observing actions change from experience, and thus deducing what someone *must* have learned, regardless of what he says. The neural record describing *how* to braid must be stored differently from the event memory of the instruction sessions this patient could not remember receiving. If we are willing to check self-report at the door, then we may enter the realm of the brain's shadow learning system.

While *explicit memory* serves itself up for conscious reflection, *implicit memory* does not. That is why it escapes our notice. The gulf between learning and awareness in the would-be weaver gapes just as widely in the healthy brain. All of us acquire wonderfully complicated knowledge that we cannot describe, explain, or recognize.

Consider the following study: the researchers Barbara Knowlton, Jennifer Mangels, and Larry Squire gave subjects the task of predicting the weather in a simple computer model. On each trial, a computer screen showed one, two, or three of the cues pictured on the next page. The subject's job was to predict whether such hints combined to herald rain or shine in the computer's phantasmagorical world. Each subject looked at the cues and typed in his answer, and the computer gave feedback, telling him whether his meteorological prognostication had been right or wrong. Then he tried again.

The researchers designed this experiment so that the displayed cues, as unhelpful as they look, *did* relate lawfully to the ultimate outcome of showers or sun. The relationship between cues and effect, however, was a complex, probabilistic function that even the smartest person couldn't deduce. By deliberately making the task

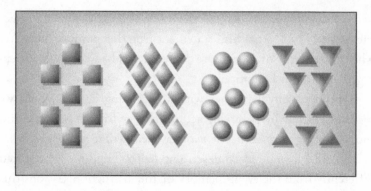

Clues given to subjects attempting to predict the weather.
(Adapted from Knowlton et al., 1996.)

too difficult for logic to unravel, the investigators hoped to neu-
tralize neocortical reasoning—so that subjects would confront the
task with one brain tied behind their backs, so to speak. The cognitive obfuscation succeeded; none of the subjects figured out the
scheme whereby clues pointed the way to a weather prediction. Despite their incomprehension, subjects nevertheless steadily honed
their forecasting abilities. After just fifty trials, the average subject
was right 70 percent of the time. Even though subjects didn't understand what they were doing or why it worked, they were still
able to *do* it. They gradually developed a feel for the situation, and
intuitively grasped the essence of a complex problem that their
logical brains could not crack.

Predicting a day's *real* weather involves a different set of clues
but, until the advent of modern meteorology, relied on the same
process that this study explores. The crisp, empty blue of a morning sky, the direction of wind, a hint of coolness to the air, an in-
definable smell just out of reach (and perhaps, for some, a twinge
of rheumatism in the knee) can combine to sketch a foreglimpse of
afternoon rain or snow by nightfall. "It *feels* like rain today," one
thinks, while looking at a sky obstinately clear of clouds.

For the amateur predictor (as most of us are), this inner, so easily ignored sense may be the best available guide. In tasks similar to weather prediction, one study found that conscious attempts at problem-solving got in the way of burgeoning intuition and actually impaired subjects' performance. Another experiment demonstrated that carefully explaining the significance of the clues in advance improved how well subjects *understood* the task, but not how well they *did* it.

Knowlton, Squire, and Seth Ramus tested the limits of implicit memory by inventing a novel grammatical structure—a set of complicated, arbitrary steps for assembling "words" in an original and utterly useless language.

This scheme can generate an infinite number of different words using T, V, J, and X, but not all combinations are valid. (XXVXJ is an acceptable word, for instance, but TVXJ is not.) The researchers did not divulge their complex rule system. They simply offered a list of fifty authentic words, and then asked subjects to judge the legitimacy of candidate words they had not previously seen.

Experimenters found that people were able to distinguish new words that adhered to the artificial grammar from those that violated

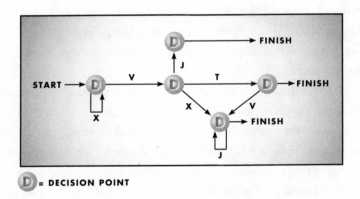

The "word"-generating scheme in an artificial grammar study.
(Adapted from Knowlton et al., 1992.)

GRAMMATICAL	NON GRAMMATICAL
X X V T	T V T
X X V X J J	T X X X V T
V X J J	V X X X V J
V T V	V J T V T X

Grammatical and nongrammatical "words" in an artificial grammar.
(Adapted from Knowlton et al., 1992.)

it. What the subjects *couldn't* do was specify how they were reaching their correct determinations. Once again, they had mastered the inner workings of an intricate system in a way that they could not render specific. They could only say they were using their intuition.

How could people make these complex judgments absent any understanding of the basis? A part of the brain must apprehend the artificial grammar's elaborate pattern without involving the neural systems responsible for comprehension. That brain mechanism has to be implicit memory, because patients with hippocampal damage (and no explicit memory) performed the artificial grammar and weather prediction tasks as well as normal subjects did. The brain's dual memory systems stand as perfect complements: damage to the organs of implicit memory leaves behind normal learning for events, facts, and lists, but obliterates the silent, incremental acquisition of intuitive knowledge human beings depend on.

The scientific study of intuition is just beginning. Researchers are already probing its power. In a 1997 study of uncommon elegance, Antoine Bechara, Hanna and Antonio Damasio, and Daniel Tranel gave people $2,000 in play money and four decks of cards to choose from. Subjects did not know the decks were rigged: turning over a card paid $100 in two decks and $50 in the other two. As in life, the high-paying decks also contained high-penalty cards, where the low-paying decks entailed fewer and smaller fines. Overall, playing exclusively from $50 decks was the winning strategy, but subjects knew nothing of the cards upon first encountering them.

After suffering a few big losses, people began to show tiny elevations in sweating as they considered drawing a card from the risky decks. Bodily tension was the only indicator of an impending hunch; by the twentieth round none could express any verbal inkling that half the cards were stacked against them. After fifty or so turns, people began to suspect that they should avoid the $100 decks, although they couldn't explain the reason for doing so. After playing eighty cards, two thirds of the subjects had figured out which decks to choose and why. Although the remaining third did not attain this conceptual stage, they were still able to win by using their sharpening intuition.

As we move through the world we tend to presume that success comes from understanding. The brightness of rationality's narrow beam makes this supposition nearly inescapable. "Reason is the substance of the universe," Hegel crowed in an age when science still expected to explicate everything. But these memory studies have intuition leading comprehension by a country mile; they reveal our lives lit by the diffuse glow of a second sun we never see. When confronted with repetitive experiences, the brain unconsciously extracts the rules that underlie them. We experience the perceptible portion of this facility as a gathering pressure in the solar plexus, ready for use but defying description. Such knowledge develops with languorous ease and inevitability, stubbornly inexpressible, never destined for translation into words.

Aristotle drew the distinction between knowing *that* something is so and knowing *why*. The restless desire of the Athenians to seek causes marked the first unsteady steps of scientific exploration. Their explanations have metamorphosed into myth, but their hierarchy of knowing endured: real knowledge, true knowledge, comes from knowing *why*. The medieval definition of *scientia* was just that: *cognitio per causas*, knowing the cause. The science of our day is confirming the utility, even the supremacy, of knowing *that* X is so without *why*. Comprehension's proper role is icing on the cognitive cake. Reason, as Pascal observed, is the slow and tortuous method by which those who do not know the truth discover it.

Implicit memory ensures that camouflaged learning permeates our lives. Spoken language, for instance, is based on a labyrinthine array of phonological and grammatical rules that native speakers know but could not explicate; most could not even recognize the rules when spelled out in plain English. We can tell instantly, as Steven Pinker observes in *The Language Instinct*, that *thole, plast,* and *flitch* are not English words but they could be, where *vlas, ptak,* and *nyip* cannot be English. Most of us have no idea why this superficially capricious distinction prevails—the quixotic "Why *not* nyip?" is immediately suppressed by its indisputable feeling of foreignness. "My brother can be died," as Pinker notes, will grate on the mind's ear, despite its coexistence with permissible homologues "My ball can be bounced" and "My horse can be raced." Only a handful of grammarians can explain why. Implicit knowledge makes language structures available for automatic use but not reflection. Children learn to speak without instruction; they absorb linguistic rules as a sponge absorbs water. Every language is intricate, but none is chaotic; the underlying uniformities reveal themselves to the neural systems poised to pluck recurring patterns out of a sea of experience.

Behind the familiar bright, analytic engine of consciousness is a shadow of silent strength, spinning dazzlingly complicated life into automatic actions, convictions without intellect, and hunches whose reasons follow later or not at all. It is this darker system that guides our choices in love.

Denise Levertov again:

Look inward: see me
with embryo wings, one
feathered in soot, the other

blazing ciliations of ember, pale
flare-pinions. Well—

could I go
on one wing,

the white one?

Fly into love's country on light and no shadow? Not a chance, as we shall see.

A CHILD'S GARDEN OF MEMORY

A child's dual memory banks mature at different rates. The structures generating explicit memories are immature at birth, and they require years of neurodevelopment to become fully functional. Implicit memory needs no warm-up; it is operational before a baby is born. In later life, the explicit memory system slowly degenerates as decades advance, while the implicit system retains its youthful robustness.

These separate maturational courses, like skewed lines in space, chart divergent trajectories for knowledge. Once a person sees the other side of thirty, he finds his power to retain individual bits of data on the wane. As the years pass he can expect to strain for the names of acquaintances, where he left the car keys, or sometimes the car. But his intuition endures and accrues. The brain's division of memory labor upholds the adage: you never forget how to ride a bike. People don't forget any capacity that depends on feel rather than fact. Because explicit memory doesn't work well at *either* end of a lifetime, people cannot retain event memories before the age of two. Freud declared in a letter of January 24, 1897, to his colleague Wilhelm Fliess, that he had retrieved a patient's memory from eleven months of age, allowing him to "hear again the words that were exchanged between two adults at that time! It is as though it comes from a phonograph." Freud's brilliance notwithstanding, he is claiming for his patient a recollective precocity that outstrips Mozart's musical one.

If infants are not recording autobiographical incidents, what are they learning? Because a baby has minimal motor control, he cannot readily demonstrate his prowess, but several clever experiments have affirmed that infants are formidable students. By monitoring infants' physiologic reactions to novelty, researchers can determine which events elicit nothing more than the autonomic equivalent of a yawn—and thus they can tell what's new and what's known to an infant's mind.

These techniques prove that babies remember their mothers' voice and face within thirty-six hours of birth. Within days, an infant recognizes and prefers not only his mother's voice but also her native language, even when spoken by a stranger. You might think this knowledge comes from postpartum interactions—quick learning indeed. But a newborn doesn't recognize his father's voice, indicating that neonatal preferences reflect learning *before* birth. The auditory system's rapid development *in utero* and the watery womb's excellent sound system surround the fetus in a symphony. Bathed for nine months in his mother's vocalizations, a baby's brain begins to decode and store them—not just the speaker's tone, but her language patterns. Once born, a baby orients to the familiar sounds of his mother's voice and her mother tongue, and favors them over any other. In doing so, he demonstrates the nascent traces of both attachment and memory.

Like the acquisition of spoken language, emotional learning happens implicitly. Even with the help of his intrauterine head start, a child takes many months to comprehend full sentences and a bit longer to produce them. Facial expressions, tone of voice, and touch carry a mammal's emotional messages; as we saw in chapter 3, a baby is born fluent in that signaling system. Implicit memory is the brain's sole learning component in the first years of life, when mother and child are bound together through their limbic connection.

Can emotional memories be recorded outside the explicit system? Antonio Damasio has shown they can. Damasio's patient Boswell is like Mr. Underwood—he has no explicit memory. Damasio and Daniel Tranel observed, however, that Boswell's bonding behavior was not diffusely random; he seemed particularly attached to a certain nursing aide. Intrigued by his affectionate selectivity, Tranel and Damasio designed an experiment to test Boswell's ability to form and preserve emotional memories. The investigators briefed three coconspirators on how to act: "Good Guy" was flattering and solicitous to Boswell, "Neutral Guy" reserved and bland, and "Bad Guy" downright unpleasant. Boswell later truthfully denied memory of meeting any of them—that knowledge never found its way into his long-term storage. But when forced to choose whom he would ask for gum or cigarettes, he stuck with Good Guy more often than chance predicted. Without event memory, without the ability to remember a name or a face, Boswell retained an emotional impression.

The poet Charles Baudelaire once wrote that the devil's finest trick is convincing the world he doesn't exist. Implicit memory has done the same. Ask someone about his emotional memory, and he'll begin by recounting the jarring discontinuities that catch his retrospective eye—the faithful dog hit by a car when he was five, the family's move from Bakersfield to Boston when he was nine, the high school prom made miserable by a rebuff from the raven-haired beauty his heart was set upon. What could be more natural than assuming that the traumas that tower in memory have affected us most? Some of them do leave their mark, but the slow and surreptitious implicit system is the true scribe of emotional learning.

A child enveloped in a particular style of relatedness learns its special intricacies and particular rhythms, as he distills a string of instances into the simpler tenets they exemplify. As he does so, he arrives at an intuitive knowledge of love that forever evades consciousness. He owes the ignorance of his own heart not to repression but to the brain's dual memory design. The frustrating illegibility of love's book is, as software makers say of problems with their programs, a feature and not a bug.

The brain's implacable condensation of precepts is its strength and its downfall. Implicit memory extracts a principle for the same reason that Mallory scaled Everest—"because it is there." Encountering an early series of consistent instances can implant an erroneous generality in a child's mind. This mental machinery distills and does not evaluate; it cannot detect whether the larger world runs in accordance with the scheme it has drawn forth from the emotional microcosm of a family. Just as grammatical English emerges from our lips automatically, a structured pattern of emotional relatedness emanates from each of us.

We play out our unconscious knowledge in every unthinking move we make in the dance of loving. If a child has the right parents, he learns the right principles—that love means protection, caretaking, loyalty, sacrifice. He comes to know it not because he is told, but because his brain automatically narrows crowded confusion into a few regular prototypes. If he has emotionally unhealthy parents, a child unwittingly memorizes the precise lesson of their troubled relationship: that love is suffocation, that anger is terrifying, that dependence is humiliating, or one of a million other crippling variations. Tolstoy was right: happy families are blandly similar (much in the way that healthy bodies are), and unhappy families unique in the exact and varied configurations of their pathology.

Take, for instance, a young man unhappily single with good reason. For as long as he can remember, his romances travel the same track. First, the shock of love with its vertiginous rush and the sweet fire in his spine. Mad mutual devotion follows for weeks. Then the first alarming note: a trickle of criticism from his partner. As their relationship settles in, the trickle becomes a torrent and the torrent a cataract. He is lazy; he is thoughtless; his taste in restaurants is banal and his housekeeping habits a horror. When he can't stand it any longer, he breaks off the relationship. Blessed silence and relief descend. As the weeks drag by into months, his newfound ease slides over into loneliness. The next woman he dates reveals herself (after a brief time) to be the doppelgänger of his recently departed ex. Without a woman, his life is empty; with her, it's misery.

These incessant cycles are the present-day echoes of a primal duet, a long-remembered melody from implicit memory. Taken together, his girlfriends present the sketch his mind recorded of his mother—an intelligent and creative woman, but with a short temper and a tendency to externalize and blame. His young brain absorbed *that* equation; he expects to find *that* archetype wherever people love. For reasons we will outline in the next chapter, he cannot find anything else. Left to himself, he will not realize there is something else to be found.

Even common sense will mislead him. He might think, as many reasonable people do, that analyzing childhood pivot points will resolve his troubles. The simplicity of this supposition appeals to both patients and psychotherapists—how easy for us to believe in a single, concentrated cause for complexity, and how hard to find visceral satisfaction in the accretion of infinitesimal influence that is more often nature's way. Hot on the trail of fugitive Mnemosyne, a therapist may root around in the patient's explicit past as he might a musty kitchen cabinet, looking to "uncover" the event memories of unpleasantness that, once wrenched into the daylight, can deliver redemption.

Turning psychotherapy into a treasure hunt for the explicit past is misguided. Exposure to a style of relatedness imprints a person with its grammar and syntax. The perceptive observer can see the stamp of that knowledge everywhere: in dreams, work, relationships, in the way he loves his wife, his children, and his dog *today*. Autobiographical memories are useful, because the pattern gleams there, too, often stark and unadorned. But explicit memory is not a shrine. Every day the patient parades the jewels of memory that the therapist seeks; they are woven inextricably into the tapestry of his life. He can no more leave the mark of his past behind than he can his face or his fingerprints. All are visible to someone who looks in the right place.

People rely on intelligence to solve problems, and they are naturally baffled when comprehension proves impotent to effect emotional change. To the neocortical brain, rich in the power of abstractions, understanding makes all the difference, but it doesn't count for much in the neural systems that evolved before understanding existed. Ideas bounce like so many peas off the sturdy incomprehension of the limbic and reptilian brains. The dogged implicitness of emotional knowledge, its relentless unreasoning force, prevents logic from granting salvation just as it precludes self-help books from helping. The sheer volume and variety of self-help paraphernalia testify at once to the vastness of the appetite they address and their inability to satisfy it.

REALITY FLIGHTS

Implicit memory warps our window on the world—one of many mental mechanisms that do so. The brain never permits naked reality to intrude into consciousness; all inbound sensory impressions pass through a process that sands the rough edges off an inhospitably complex universe. For a demonstration, close an eye, push gently on the corner of the other—and the world dips or lurches several degrees, as if it were not your fingertip moving a millimeter but the hand of God shaking the planet. The brain doesn't detect the eye's position but tracks only the ocular movements it commands. When it orders no eye movements, the brain assumes none occurred—accurate in every situation save one. Displacing the eye manually shifts the light falling on the retina. The brain concludes that with eyes immobile, the world has turned. All experience comes to us through similar layers of invisible and occasionally dubious deductions.

But like the Wizard of Oz, your brain encourages you to pay no attention to the man behind the curtain. The retina registers color only in the central thirty degrees of the visual field. Visual virtuality, on the other hand, is an omnidirectional chromatic canvas—with many hues inferred and presumed and painted in for our viewing enjoyment. Assuming the world *is* the way it *looks* is the neurally prompted so-called naïve realism to which most of us unwittingly subscribe. Among the many certainties in life, Umberto Eco writes, one is supreme: "All things appear to us as they appear to us, and it is impossible for them to appear otherwise."

Our internal realities are mock-ups of unparalleled persuasive power. The tangle of neurons that make up a person, after all, are the same ones that generate the disparity between reality and experience. Of course, glitches do occur. If a virtual world misrepresents reality only slightly, we call that an illusion; if the discord is substantial, a hallucination. Psychosis is a sweeping and catastrophic disparity between the individually virtual and the clean, hard edge of the veridical. Even digestive sensations have their own replica in the brain, thus enabling strange illnesses—like that of a woman who, after a stroke, felt the food she swallowed travel down her throat and descend into a nonexistent cavity in her left arm, a disquieting disruption in visceral virtuality. The limbic brain, too, models the world, making our emotional realities a set of neurally generated phantoms loose in the mind.

Reality is thus more personal than daily life suggests. Nobody inhabits the same emotional realm. Many people live in a world so singular that what they see when they open their eyes in the morning may be unfathomable to the rest of humanity. When one woman looks at an attractive man, she sees someone who wants to possess her and stifle her creativity; another sees a lonely soul who needs mothering and is crying out for her to do it; a third sees a playboy who must be seduced away from his desirable and unworthy mistress. Every one of them knows what she sees and never doubts the identity of the man in front of her faithful retinas, her fanciful brain. Because people trust their senses, each believes in her own virtuality with a sectarian's fervor.

It's the rare person who glimpses the expanse of his own subjectivity, who knows that everything before his mind's eye is the Hindu's *maya*—an elaborate dream of the world worthy of a god, but reverie just the same. Only a person of surpassing wisdom doubts his own mind enough to remark, as the Supreme Court justice Robert Jackson once did when reversing himself on a point of law, "The matter does not appear to appear to me now as it appears to have appeared to me then."

Six

A BEND IN THE ROAD

HOW LOVE CHANGES WHO WE ARE
AND WHO WE CAN BECOME

It is 7:15 and twilight is gathering around the hospital, when a man enters the emergency room. He is fiftyish, with a paunch to his belly and a grayish tint to his face. He has chest pain, he tells them at the front desk, a gnawing, persistent discomfort behind his breastbone. The triage nurse charts his vital signs: heart rate, blood pressure, and respiratory rate, all elevated. He dons the obligatory paper gown and takes his place on a gurney as an intern readies the cardiac monitor.

If a wayward clot or a shard of crystalline cholesterol has blocked a cardiac artery, part of his heart will die, and he will need admission to the hospital, where the staff and their mechanical aides will try to save his life. But as is so often the case in medicine, his symptoms do not point in one direction: they fit with dozens of ailments, most comparatively harmless—an ulcer, an anxiety attack, a pulled muscle, an undigested bit of beef. If everyone with his chest pain were admitted, our health care system would reach bankruptcy sooner than it will on its present course. No diagnostic test is definitive here; the situation calls for expert judgment applied to stubbornly indeterminate facts. Thousands of times a day, armed with less information than can lead to certainty, someone decides: is this person having a heart attack?

Physicians have scrutinized this question for centuries, gathering the slender hints and occult signs whereby an ailing heart betrays its presence. They understand what a heart attack is, how and where and why it occurs, whom the risk factors prey upon most avidly. Still, medical skill at detecting myocardial infarction is far from perfect. William Baxt in the Department of Emergency Medicine at the University of California, San Diego, hoped to augment human diagnostic proficiency by enlisting a computer assistant capable of learning to distinguish heart attacks from the varied conditions they resemble. Baxt fed the program details from the case histories of 356 patients. In the next 320 chest pain cases, physicians made the right call four fifths of the time. The computer scored 97 percent.

The *de facto* division of labor between man and machine usually splits along pleasing lines. Computers typically excel at the tireless repetition of algorithms—the menial mental labor that human beings have little inclination to undertake. Supercomputing inroads into championship chess rely on mechanical combinatorial might, not clever strategy. Prodigious calculators they may be, but our smartest machines can't make sense of a simile, summarize a sitcom plot, or take the dog for a walk. One wonders, then, how a few hundred lines of programming code, with no ability to understand anything about cardiac physiology, so rapidly outpaced human physicians in arriving at a medical diagnosis.

A human being has dual hearts—the first, a pulsating fist of muscle in the chest; the second, a precious cabal of communicating neurons that create feeling, longing, and love. The two hearts intersect for a moment here because the program that so brilliantly assessed cardiac endangerment is a *neural network*—poetically named, for it is neither neural nor network, but a series of mathematical statements that model the brain's own computing agility.

The *modus operandi* of the neural network is unique. Standard computer software runs on the expertise, routines, and contingent responses that human masterminds script in advance. Such a program cannot handle any situation that the writer cannot foresee; once written, the program itself does not change. The meticulously crafted core of a neural network, on the other hand, learns from experience and transforms itself—a capacity created by small-scale software modeling of the communication that occurs among organic, brain-dwelling cells. Before a neural network can compute a solution, it first absorbs information from intensive training sessions. That learning phase gradually alters the program's innards. Neural networks (also called *parallel-distributed processing* or *connectionist* models) excel at gleaning subtle patterns that hundreds of variables jointly determine. The best neural networks are more astute at diagnosis than doctors, better at forecasting weather than meteorologists, and more profitable stock pickers than mutual fund managers.

A neural network is machine intuition. After a connectionist program delivers its answer, one cannot obtain meaningful knowledge about the processing details—*why* it says Patient A suffered a heart attack but Patient B did not. Examining the network yields minimal information about the basis for its conclusions. Because a neural network taps into the brain's own data-processing mechanism, it arrives at sophisticated, unanalyzable inferences—as does humanity's emotional heart. Understand how a neural network functions, and you will know the innermost secrets of the intuitions that guide us in love.

In the previous chapter, we looked at memory from a macrocosmic perspective—how a person remembers and learns. Neural network theory begins at the opposite, miniature pole: mathematicians translate the brain's memory-making steps into equations, implement them on computers, and strive at elevating their apprentices into better learners. They have not only produced a program wise in the ways of heart attacks, but they have also radically altered how scientists think about living mechanisms of learning. And the quest has come full circle: the thriving new field of *computational neuroscience* applies the heady mathematics of connectionism back to the original problem of divining how brains—and thus human lives— operate the way they do. These discoveries reveal a last, luminous power of the limbic domain: love alters the structure of our brains.

MEMORIES ARE MADE OF . . . WHAT?

Under the right conditions, a collection of neurons can learn. A rat that runs a maze, a German shepherd that sits on command, and a child who recites the Gettysburg Address reflect the ability of a nervous system to record information and hold it in abeyance. Years may elapse before the rat, dog, or child uses that preserved data to influence the muscular contractions we know as behavior. Everyone who attended an American elementary school can remember, for instance, how the Gettysburg Address begins, the initial syllables rising to the surface of the mind and the tip of the tongue as readily as a magician might draw a chain of knotted scarves from Lincoln's stovepipe hat. Recalling this minute datum is a masterful conjuration in physiology: *that* single improbable sequence is somehow inscribed within a group of living cells. It is one mote among billions, encrypted in suspended animation, and the organic function of those cells whisks it into consciousness on a moment's notice. Lincoln's apparently unforgettable phrase remains embodied in the marvelous entanglement of the brain's cellular components for as long as they live. But how?

Any system that aims to warehouse data forms a material record. The King James Version Bible stores its truths in scattered dots on a page, "The Mona Lisa" in pigments on a canvas, a compact disc in pits on a glittering plastic platter. The mechanisms may vary, but the end is unswerving: a durable physical representation of knowledge.

The brain has no dots or paints or pits, only neurons. Every mental activity—contemplating a theorem, savoring a hot fudge sundae, dreaming of the boy next door—consists of neurons firing in a certain sequence. When neurons cache Lincoln's remarks at Gettysburg, the specific neural sparkle defining *that* string of words must be made to last. Millions of neuronal flash patterns course through the brain every minute. How does any *one* attain permanence?

MEMORY'S METAMORPHOSIS

Consider a simplified network model for storing and retrieving a sensory input. Suppose that in this array, weak tendrils join each neuron to every other:

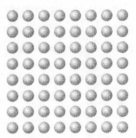

This network is naïve; it stands ready to receive experience but has recorded none.

A sensory input arrives:

A pianist transforms the cryptic swirling of dots and lines on the music in front of him into the fluttering of his fingers, and then into beauty. Here the reverse occurs: the richness of a sensory experience is translated into the brain's peculiar notation system— no staffs or semiquavers; instead, neurons fire.

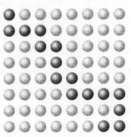

One needn't linger over the details of that conversion—why *this* neuron and not *that* one? Somewhere, an array of neurons depicting sensory data about the stimulus will flicker into brief life. A brain needs millions of neurons to portray such a symbol, but this example will proceed more comfortably on a smaller scale—in this neural network, registering those crisscrossed lines necessitates just sixteen. To make a memory, the network immortalizes the association of that *particular* group by strengthening their previously faint linkages.

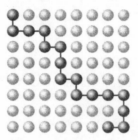

When the figure passes from view and the cells quiet down, the skeletal remnants remain:

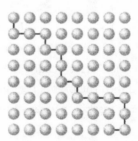

The fortified conjunctions permit these neurons to fire together again. When a few go off (A), they trigger their quiescent fellows along the slightly worn paths between them. Like a string of dominoes, they race to tip each other to a common fate. The old pattern is rejuvenated (B), and with it, a recapitulation of the original character.

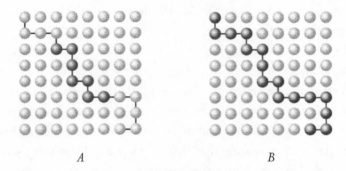

A   B

This storage scheme, the brainchild of psychologist Donald Hebb, is a powerhouse. Hebb proposed the mechanism a few years after World War II drew to a close. Only within the past fifteen years, however, did researchers explore its mathematical premises and build large-scale computer models of Hebbian learning. Both endeavors—the mathematical insights and their implementation in computer simulations—have illuminated more than a few of the mysteries about why people think and feel the way they do.

Hebb's central proposal remained theory until the advent of experimental techniques for taking electrical measurements from individual brain cells. In a refreshing physical affirmation of mathematical abstraction, the data demonstrate that neurons in the living brain behave as Hebb predicted. The brain makes memories by enhancing the couplings between concurrently firing neurons.

A printed page in this book may survive for hundreds of years. CDs have a shelf life of one or two decades and the average Etch A Sketch scrawl lives minutes at most. A neuron's "on" state lasts for a thousandth of a second. The ephemeral span of its signals forces the brain to render present and past data differently. At any instant, the precise configuration of firing neurons specifies what a brain is representing *now*. But the past lies dormant within the network's structure, formed by accumulated links of varying potency. Each constellation of mute connections embodies the potential for a previous ensemble to be reanimated and remembered. This sepa-

ration of past and present makes the network into a living and eccentric time machine.

Every bit of life impinging on the brain changes some of its
links, although any individual datum affects only a minuscule fraction of the innumerable totality. As subtle changes accrue, experience rewires the microscopic structure of the brain—transforming
us from who we *were* into who we *are*. At a Lilliputian level, the brain
is an elaborate transducer that changes a stream of incoming sensation into silently evolving neural structures. Minor events exert only
a transitory alteration in a few far-flung neuronal ties, while formative experiences lay down resilient patterns that prevail for a lifetime.

And the limbic brain, wherein some of those cryptographic patterns reside, can reach beyond the frail borders of one mind and
into another—a fact with far-reaching consequences.

THE ECHO'S RING

Construct a neural network and set it in motion, and its odd memory mechanism begins to work divisive magic. Neurons that fire together once tend to do so again as they become bound to one
another with increasingly close ties. Cells that are never on simultaneously start to suppress one another. The once homogeneous
network spontaneously fractures into squabbling cliques. Team
members spark one another to fire *en masse.* Opposing squadrons
fight for the chance to be active. At any moment, a single neuron
receives stimulation from compatriots and inhibitory signals from
enemies. As the cells trade their flurry of pros and cons, each finds
at its own activity level. The network as a whole then settles into a
certain conformation of active and inactive units. A network is
most stable when a set of allies is firing—when one team wins.

In the brain, some neurons receive input from ten thousand others and may deliver outgoing messages to ten thousand more. With
such extensive dissemination of signals, a simplifying assumption
of total commingling is not far off the mark. Just as team member

cells goad one another into excitability, so do compatible *networks* motivate one another in the brain. Likewise, dissonant networks compete and drive one another down. The unforeseeable result of this catholic discourse is that emotional memory skirts the linear flow of time.

When a network turns on, it immediately dispenses electrical encouragement to every other concordant network. And the secondaries light up in direct proportion to their shared affinity. If network A fires, and B is highly compatible, then A will galvanize B. This companionable coactivation echoes on: with B aroused, *its* allies will awaken, and so forth. Like ripples blooming from a central splash, memory networks spread out along lines of similarity, bringing the most mutually congruent to life, fading in influence as correspondence drops off.

Think the word *dog,* and the circuits encoding for *German shepherd* and *golden retriever* warm up in your mind, and those for *walk* and *bone* and *flea* a little less so. The strong activation of *dog* leaves *mutual fund* in hibernation (except in the unlikely event of an idiosyncratic bridge—today Fido ate your brokerage statements, say). Computer-based neural networks operate this way, and so do human beings. Showing a person the word *dog* actually makes him respond more quickly to words like *bone* and *flea,* while reaction time to *mutual fund* remains unchanged.

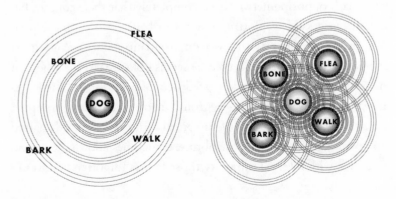

For those animals with a limbic brain, emotionality forms a principal dimension of that associative network. In place of *dog* retrieving *bone* and *walk* and *flea,* a particular emotion revives all memories of its prior instantiations. Every feeling (after the first) is a multilayered experience, only partly reflecting the present, sensory world.

In chapter 3, we saw that the evanescence of emotions, their pulse-and-fade propensity, is nearly musical. Now the metaphor draws closer. A musical tone makes physical objects vibrate at its frequency, the phenomenon of *sympathetic reverberation.* A soprano breaks a wineglass with the right note as she makes unbending glass quiver along with her voice. Emotional tones in the brain establish a living harmony with the past in a similar way. The brain is not composed of string, and there are no oscillating fibers within the cranium. But in the nervous system, information echoes down the filaments that join harmonious neural networks. When an emotional chord is struck, it stirs to life past memories of the same feeling.

One manifestation of these orchestral evocations is the immediate selectivity of emotional memory. Gleeful people automatically remember happy times, while a depressed person effortlessly recalls incidents of loss, desertion, and despair. Anxious people dwell on past threats; paranoia instills a retrospective preoccupation with situations of persecution. If an emotion is sufficiently powerful, it can quash opposing networks so completely that their content becomes inaccessible—blotting out discordant sections of the past. Within the confines of that person's virtuality, those events didn't happen. To an outside observer, he seems oblivious to the whole of his own history. Severely depressed people can "forget" their former, happier lives, and may vigorously deny them when prompted by well-meaning guardians of historical verity. Rage affords hatred an upper hand that is likewise obtuse, sometimes allowing a person to attack with internal impunity those he has forgotten he loves.

The consequences of emotional reverberation in the brain's networks reach beyond selective amnesia during a dominant mood. A childhood replete with suffering lingers in the mind as bitter, encoded traces of pain. Even a tangential reminder of that suffering can spur the outbreak of unpleasant thoughts, feelings, anticipations. As if he had bumped a sleeping guard dog, the adult who was an abused child may feel the fearsome jaws of memory close after he glimpses a mere intimation of his former circumstance. In a sad empirical confirmation, maltreated children flipping through pictures of faces exhibit a hugely amplified brain wave when they encounter an angry expression.

Other people are troubled by emotional-memory networks that are simply too ready to pass around the signals that comprise negative feelings. Such a person finds he can't shake an unpleasant emotion once it gets going. Rather than dwindling within minutes as they should, an emotion and its associated repercussions may drown out the rest of his mind for days. That kind of limbic sensitivity makes the thousand natural shocks the flesh is heir to well-nigh unbearable.

Remedies do exist for those whose networks engage in excessive emotional reverberation: some psychopharmacologic agents act as the damper pedal on a piano does, applying a gentle, restraining influence on buzzing strands. Why they possess this property is not yet known, but the impact of these medications on emotional virtuality is what one might expect: emotional chords are quieter and fade sooner. For those whose limbic networks are high-strung, the relief can be lifesaving.

One person, for instance, related that every minor setback made her ruminate for days. "I know it doesn't make any sense," she said, "but my boss corrected my spelling on a report the other day, and my mind wouldn't stop: he thinks I'm incompetent, my work isn't good enough, I'll lose my job. All of that is ridiculous, I know. I'm the best manager he's got. But when anything goes wrong, I just can't shake this awful feeling." So prominent was her sensitivity to emotional slights that she retreated from intimacy. No matter how cautious, her partner was bound to do or say something that hurt her feelings, and then she felt terrible for weeks. Being in a relationship, she said, was like trying to dance barefoot—eventually her toes would get bruised and she would flee.

A touch of the right medication diminished her emotional twanging to a normative range. For the first time in her life, she was able to feel a minor pang. She could be upset for half an hour or so, and then get on with her day. "Is this what life is like for everyone else?" she asked. "No wonder they can be in relationships." With her vulnerability reduced to livable levels, she was readier for love. As she said, she was now dancing with shoes on.

THE ROAD CURVES

As a neural network sees more of the world, its ensuing quirks and kinks confound and complicate the human experience of love.

Here is the network after encountering the first sensory input.

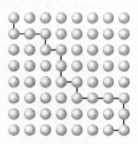

When a second, similar sight presents itself—

—the network again disassembles those features and transmutes them into an assortment of winking neurons. Because Instance Two resembles Instance One, this neuronal rendition overlaps considerably with the last.

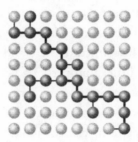

Hebbian machinery reinforces these connections:

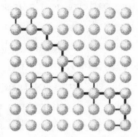

If we show it a third and then a fourth similar item,

the process repeats: these items, too, utilize many of the same neurons, and the links between them grow ever stronger.

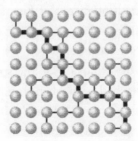

When the network rests after storing the four information sets, the curious aspects of neural memory reveal themselves.

For starters, memory within the brain is fluid. Like a squirrel dispersing nuts to multiple hiding places to thwart theft, the brain scatters its memory treasures across a number of individual connections. The boon of distribution is security; the disadvantage, infidelity. A brain can lose a neuron here and there and the stored data suffers relatively little, a property of neural network memory termed *graceful degradation.* But as new facts rain down upon and trickle through the network, some older links are dissolved. Dissimilar information patterns can cohabitate in the brain without much mutual disturbance, because they rely on largely different sets of neurons. But like patterns jostle and coincide and overlie, and cannot avoid erasing and emending each other's contours.

The brain's past is not carved into the solid rock we would like (and imagine) memory to be. Instead, the story of each life is traced on a sand dune that the winds of time and experience gradually sculpt from one shape to another. After their first instant, our memory traces start shifting away from what they were, and we can never retrieve the pristine data they once encoded.

One inconvenient outcome of this mnemonic drift is the miserable fallibility of eyewitness testimony. Although most people do not realize it, they are incapable of remembering events as they happened. In study after study, they incorporate fragments from earlier or later incidents, general expectations and implications gleaned from leading questions into their memories of what they saw and heard. The resulting ragout of fact, fancy, suggestion, and innuendo feels as convincing as a more accurate memory would. Psychologist Ulric Neisser interviewed forty-four students the morning after the 1986 explosion of the space shuttle *Challenger,* asking them where they had been when they first heard of the disaster. He repeated the questions two and a half years later. None of the later accounts correlated entirely with the originals, and fully a third of them were "wildly inaccurate." The students were cheerfully ignorant of the fabrication that passed in their minds for a memory; many insisted that the recent, wrong versions were actually right. "As far as we can tell," Neisser observed, "the original memories are just gone."

In a neural network, new experiences blur the outlines of older ones. The reverse is also true: the neural past interferes with the present. Experience methodically rewires the brain, and the nature of what it *has* seen dictates what it *can* see.

In our network, a group of neurons stands out by virtue of joint ties: they share connections that the network strengthened three or four times.

This group of cumulative links stores the elements that the four inputs have in common. Blend the quartet together, and their content is manifest:

This composite preserves shared features—dual vertical pillars united by a single horizontal crossbar. Individual elements, such as serifs and swirls, register faintly and tend to wash out. While each of the four contributors is a single, serial data point, the network encodes most strongly their synthetic summary, the *trend* they exemplify—a roman H. Simply by encountering a series of calligraphic instances, the neural network has automatically extracted and emphasized the H-ness latent within all of them.

Prototype extraction—the distillation of pure, intoxicating principles from the muddle of diverse experience—is the natural, inescapable outcome of neural memory. We have distinct linguistic labels for "memory" and "condensation," but within the brain they are one. The structure of human thought is shaped accordingly.

Here is the network with the ingrained prototype:

Show it a fifth figure:

This one is seductively similar to the prototype. And when the neural network tries to process the latest input, time spins backward and veracity breaks down.

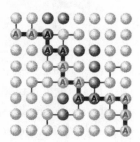

The fifth figure triggers several prototype neurons, but in a flash, *their* mutual chain reaction ignites. Like a bank of stadium lights, the prototype team blazes into phosphorescent glory.

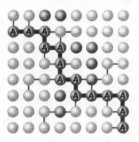

The A-team easily outshines the neurons that for one mini-moment encoded the true appearance of the last character. With the resurrection of the A-neuron band, the network departs reality. What it now portrays is a cross between factual present and proto-type past:

As the network views it, the last input *is* an H—the listing twin towers and drooping crossbar have almost assumed the customary crisp relationship that the prototype displays. These interpretive changes to the actual, ambiguous sensory input are made courtesy of the peculiar memory mechanism that a living brain is obliged to employ.

The prototype team constitutes an *Attractor*—a coterie of in-grained links that can overwhelm weaker information patterns. If incoming sensory data provoke a quorum of the Attractor's units, they will trigger their teammates, who flare to brilliant life. An Attractor can overpower other units so thoroughly that the network registers chiefly the incandescence of the Attractor, even though the fading, firefly traces of another pattern initially glimmered there. A network then registers novel sensory information *as if it conformed to past experience.* In much the same way, our sun's blinding glare washes countless dimmer stars from the midday sky.

Neural Attractors allow human beings to decipher handwriting, wherein the misshapen letters in the sensory stimulus deviate markedly from the straight and true strokes that first-graders strive to imitate and adults have long since abandoned. Translating a hand-written note into English letters is the work of a few milliseconds for a neural network but arduous labor for a computing device that processes reality *as it is,* without first subjecting it to the distorting and defining influence of Attractors. Conversely, Attractors make the act of proofreading exceptionally difficult for neural processors. Reading "taht" in the middle of a sentence, for instance, often activates the brain's heavily engrained Attractor for "that." Most of the time, the mind's inner eye will see only the autocorrected "that," thus circumventing the opportunity to correct a textual error that consciousness never experienced.

Verify for yourself that reality-distorting Attractors dwell inside your mind.

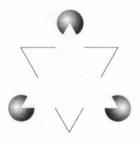

This is the Kanisza triangle, but the apparent three-point polygon is pure figment: a trio of Pac-Men, six segments, no triangle. Yet the mind's conviction of triangularity is irresistible, because these suggestive impostors conspire to activate the brain's hardwired neuronal shortcuts leading to the perception of lines and edges. Try to see only what is *really* there, and you will find the truth disappearing behind the simpler, less accurate world your brain is determined to deliver.

Or read the following:

TAE CAT

The central "letters" in both words are identical, ambiguous twins—duplicates of the fifth figure displayed to our neural network. In the first word, the brain's learned Attractor for "THE" forces the equivocal figure into an unsteady "H," while in the second, a different Attractor makes the same lines merge into an incomplete "A." Although every bit as valid from the perspective of realism, not many people will see "TAE CHT." Nobody will see the truth: "T?E C?T."

Einstein's relativity theory proved that a local concentration of mass warps space, angling the arrow flight of nearby objects, even bending the trajectory of light. The fabric of space, he said, is not a rigid plane like a billiard table or a bowling lane, impervious to the presence of the bodies traveling its surface. Instead, space is like a taut sheet of rubber indented by matter—dimpled lightly by the pea-size mass of a planet, a deep concavity stretching out from the enormous density of a sun.

The brain's habit of concentrating experience into Attractors likewise makes the mind a pliable Einsteinian fabric strewn with incurvations. At the bottom of each force field well is an Attractor, convoluting the plain, true line of thought and standing ready to

exert its influence on information patterns venturing close enough
to be twisted or trapped. Even time flows differently in the neigh-
borhood of a mass, as it does in the vicinity of a strong Attractor.

Wallace Stevens on time, heart, and mind:

It is time that beats in the breast and it is time
That batters against the mind, silent and proud,
The mind that knows it is destroyed by time.

Time is a horse that runs in the heart, a horse
Without a rider on a road at night.
The mind sits listening and hears it pass.

With time as a stallion loose in the heart, each Attractor compels
a bend in the road that horse follows. While sifting through the
sensory present, the brain triggers prior knowledge patterns, whose
suddenly reanimated vigor ricochets throughout the network. Old
information comes alive, and a person then *knows* what he used to
know. Because people are neural beings, the past is potentially vi-
brant within them.

The limbic brain contains its emotional Attractors, encoded
early in life. Primal bias then forms an integral part of the neural
systems that view the emotional world and conduct relationships.
If the early experience of a limbic network exemplifies healthy
emotional interaction, its Attractors will serve as reliable guides to
the world of workable relationships. If a diseased love presents itself to a child, his Attractors will encode it and force his adult relationships into that Procrustean bed. Because his mind comes
outfitted with Hebbian memory and limbic Attractors, a person's
emotional experience of the world may not budge, even if the
world around him changes dramatically. He may remain trapped,
as many are, within a virtuality constructed decades ago—and, as

Mark Twain observed, a person cannot depend on the eyes when imagination is out of focus.

Limbic Attractors spawn a vexing and fascinating aspect of emotional life—"transference," Freud's term for the universal human tendency to respond emotionally to certain others as if they were figures from one's past. Freud thought transference living proof that a banished memory can escape confinement and hover before a loved one's features, overshadowing a present angel with a past devil or vice versa.

Science has a way of supplanting myths with no less fantastic truths: transference exists because the brain remembers with neurons. *Any* system that undertakes Hebbian processing carries out the same distortion, whether the system is living or machine. Computer-based neural network programs are structures whose memory mechanism reduces experience into compact, occasionally fallacious expectancy. So, too, are we.

Because human beings remember with neurons, we are disposed to see more of what we have already seen, hear anew what we have heard most often, think just what we have always thought. Our minds are burdened by an informational inertia whose headlong course is not easy to slow. As a life lengthens, momentum gathers. A wistful aside from two neuroscience researchers:

> [In] scientific work, we find that new theories are understood only by the graduate students, whose intellectual identities are then wholly transformed. . . . In contrast, the senior professors are burdened with such connectional inertia that when they encounter new ideas there is no apparent effect, other than an occasional vague irritation.

WHEN WORLDS COLLIDE

No individual can think his way around his own Attractors, since they are embedded in the structure of thought. And in human beings, an Attractor's influence is not confined to its mind of origin. The limbic brain sends an Attractor's sphere of influence exploding outward with the exuberance of a nova's gassy shell. Because limbic resonance and regulation join human minds together in a continuous exchange of influential signals, every brain is part of a local network that shares information—including Attractors.

Limbic Attractors thus exert a distorting force not only within the brain that produces them, but also on the limbic networks of *others*—calling forth compatible memories, emotional states, and styles of relatedness in *them*. Through the limbic transmission of an Attractor's influence, one person can lure others into his emotional virtuality. All of us, when we engage in relatedness, fall under the gravitational influence of another's emotional world, at the same time that we are bending his emotional mind with ours. Each relationship is a binary star, a burning flux of exchanged force fields, the deep and ancient influences emanating and felt, felt and emanating.

Rachel Naomi Remen, in her book *Kitchen Table Wisdom*, describes her own brush with another mind's defining authority. As an adolescent she was ungainly, and her relationship with an older cousin centered on the woman's acceptance of her embarrassing clumsiness. When Dr. Remen matured to elegant womanhood, she could not escape her cousin's encapsulated conviction of that outgrown identity. In her cousin's presence, she reverted to tripping on curbs, dribbling food on her clothes, spilling the contents of her purse across the floor of a restaurant. The inner conception we carry of others "may be reflected back to them in our presence and may affect them in ways we do not fully understand," she writes. "Over the years, I have come to wonder if it may even be communicated more directly, by the sharing of a private image in a mysterious yet tangible way, as my cousin did with me."

The limbic transmission of Attractors renders personal identity partially malleable—the specific people to whom we are attached provoke a portion of our everyday neural activity. In the vistas of imagination, the self is a proud ship of state—subject to the winds and tides of circumstance, certainly, but bristling with masts and spars and beams, fairly bursting with solidity. We would scarcely imagine that identity could be as fluid as the seas that supposed self rides upon.

E. E. Cummings paints a lover's power to render identity in this way:

your homecoming will be my homecoming—

my selves go with you,only i remain;
a shadow phantom effigy or seeming

(an almost someone always who's noone)

a noone who,till their and your returning,
spends the forever of his loneliness
dreaming their eyes have opened to your morning

feeling their stars have risen through your skies . . .

The reach of limbic Attractors stretches beyond the moment. The *sine qua non* of a neural network is its penchant for strengthening neuronal patterns in direct proportion to their use. The more often you do or think or imagine a thing, the more probable it is that your mind will revisit its prior stopping point. When the circuits are sufficiently well worn such that thoughts fly down them with little friction or resistance, that mental path has become a part of you—it is now a habit of speech, thought, action, attitude. Ongoing exposure to one person's Attractors does not merely activate neural patterns in another—it also strengthens them. Long-standing togetherness writes permanent changes into a brain's open book.

In a relationship, one mind revises another; one heart changes its partner. This astounding legacy of our combined status as mammals and neural beings is *limbic revision:* the power to remodel the emotional parts of the people we love, as our Attractors activate certain limbic pathways, and the brain's inexorable memory mechanism reinforces them.

Who we are and who we become depends, in part, on whom we love.

Seven

THE BOOK OF LIFE

How love forms, guides, and alters
a child's emotional mind

When Dr. John Watson moves into the flat at 221B Baker Street, he happens to pick up a journal belonging to his new roommate. An article boldly entitled "The Book of Life" has been marked in the table of contents, and there Watson reads the following lines on what is called the Science of Deduction:

> *From a drop of water, a logician could infer the possibility of an At-lantic or a Niagara without having seen or heard of one or the other. So all life is a great chain, the nature of which is known whenever we are shown a single link of it. . . . By a man's finger-nails, by his coat-sleeve, by his boots, by his trouser-knees, by the callosities of his fore-finger and thumb, by his expression, by his shirtcuffs—by each of these things a man's calling is plainly revealed. That all united should fail to enlighten the competent inquirer in any case is almost incon-ceivable.*

"What ineffable twaddle!" Watson says, slapping the magazine down on the table. "I never read such rubbish in my life." The author of the piece, he discovers, is the man whose exploits he will devote his life to chronicling: the celebrated supersleuth, Sherlock Holmes.

The same sort of challenge that Holmes relished—where one is "compelled to reason backward from effects to causes"—confronts anyone who attempts to trace the emotional mind's de-velopment. The limbic detective begins not with fingernails or

trouser-knees but with discernible emotional attributes—a disposition to chronic depression, an inability to assert oneself, a lifetime spent loving inattentive partners. The task is to formulate a
historical sequence that accounts for the presence and precise conformation of those traits in one human mind.

How does a personality come into being? An infant undergoes a
startling metamorphosis as his brain develops. He starts as a wide-
eyed creature with an inborn knack for reading emotions, but before long he blossoms with elaborate emotional attributes and
skills. An identity as definite and distinct as a fingerprint takes
form, so palpable we can sense its mental ridges and whorls. When
we meet an adult we can know without too much difficulty
whether he is generous or stingy, treacherous or trustworthy, in-
timidating or obsequious. We can know if he is able to trust, to
compete, to know himself and others, to love. How does the limbic brain coalesce into a coherent structure? Beginning as diffuse
neural propensities, how does an infant become a *person*?

Holmes felt he could conduct most investigations without leaving the relaxing confines of his armchair. Freud, Holmes's contemporary and epigone, shared his conviction and may have been
tempted into emulation. Based on his reasoning skills and the clues
he derived from his seat at the head of the analytic couch, Freud
erected an immense and ornate palace of deductions about emotional development and the agencies that bring about a child's
mind. His confidence in his edifice rivaled Holmes's faith in *his* intellectual acuity.

Unfortunately, the Holmesian method of investigation is enter-
taining but impossible fiction. The sleight-of-hand artist who
draws a crisp one-hundred-dollar bill from his assistant's ear is not
minting new money; the bill is exactly where he placed it and knew
it to be all along. Likewise the enigmatic clues that Holmes so cun-
ningly pierces to Watson's dumbfounded amazement—every such

scene was crafted by a writer who knew *a priori* the identity of the criminal and the means of his nefarious activities, who knew the *one* implication his hero must snatch from millions if the plot was to proceed.

In "A Scandal in Bohemia," for instance, Holmes observes several scratches on the side of Watson's shoe. From this and this alone, he concludes that the good doctor has been tramping around in the mud, and that he must have hired a new housekeeper who scored the soles while clumsily scraping off the residue of his master's muddy perambulations. Holmes brandishes the deduction, and Watson slavishly confirms it. The two neatly bypass hundreds of equally likely permutations: Watson stepped on a rake and scratched his shoes, or marred them in cleaning them when he was drunk, or left his good shoes at the club and had to dig a worn pair out of an old and cluttered closet—and so on, almost ad infinitum.

When Holmes inspects a coat-sleeve or shirt cuff and points his unwavering finger at a murderer, he arrives at his conclusions in a fashion so simple and time-honored that one might describe it as elementary: he cheats.

The investigator of the emotional mind is denied that considerable shortcut; its Creator has not provided him with the answers in advance. A logical observer cannot start with an adult's emotional traits and forge a chain of deductions, extending backward in time to one culprit cause. Such armchair detecting is a sport that requires an impractical degree of omniscience. Instead, modern seekers must take advantage of science and the light it brings to questions about how and why a mind matures.

In the last four chapters, we recounted the different faces of the physiologic bond that unites relationship partners, including parents and children: limbic resonance, regulation, and revision. In this chapter we describe how those forces combine and conspire to create, from an infant's unadulterated vulnerability and promise, a human being and an emotional identity.

The full history of any life contains additional players whose roles we will not consider here: chance, trauma, physical illness, intellectual prowess, athleticism, poverty, race, and many more. Significant as these are in selected lives, our purpose is not to detail the influence of every factor capable of impinging on a malleable mind. Such an exhaustive (and exhausting) enterprise would, in its cacophony of contributing voices, drown rather than draw out the story of this chapter: how aspects of parental love shape a young mind.

THE BIG PICTURE

Everything a person is and everything he knows resides in the tangled thicket of his intertwined neurons. These fateful, tiny bridges number in the quadrillions, but they spring from just two sources: DNA and daily life. The genetic code calls some synapses into being, while experience engenders and modifies others.

The brain thus takes shape as a compromise between unyielding limits and nearly infinite freedom. It is like a snowflake or a sonnet, whose innumerable members remain bound to an eternal integer. The polarity of water molecules constrains a snowflake to be a six-sided polyhedron, and a sonnet comprises fourteen lines. The universe does not contain a seven-sided snowflake or a sonnet with five quatrains. But within the expanse of these restrictions lie endless permutations of beauty.

In the brain, a genetic blueprint directs the raising of rough neural *scaffolds* that serve as the cores of various subsystems. DNA thus reins in the riotous proliferation of designs that a hundred billion cells could freely generate. One brain, one plan: no brain sports three temporal lobes, and none registers anger with an upward turn to the lips. But, as in a topiary, variety flourishes within walls. Early experience trims a scaffold's semiadjustable outline into a neural *template:* an assemblage of neurons and connections fine-tuned for function in a particular environment. Genetic information lays down the brain's basic macro- and microanatomy; experience then narrows still-expansive possibilities into an outcome. Out of many, several; out of several, one.

After the template takes form, neural flexibility wanes, but often not to zero. And here the brain is like a sonnet that undergoes eternal editorial correction or a snowflake that follows a perpetual tumbling path, always adding crystals to hexagonal wings. Ongoing experience continues to mold neural connections, ensuring that one's personality never rests. As Heracleitus wrote several thousand years ago, "We do and do not step into the same river. We are, and are not."

A neural learning machine has a natural tendency to emphasize the influence of its youth. The young brain teems with far more neurons than it ultimately keeps. Most of these bloomers die out over the course of childhood as luxuriantly populated scaffolds slim down to leaner templates. Because the doomed cells and links could have stored data, their demise represents the loss of information a brain *might* have encoded. When billions of neurons depart, as they do in the brain's prolonged pruning phase, pages disappear from the mind's book forever.

Why do certain neurons survive the first years of life where multitudes perish? The life-sustaining power of attachment for mammals is mirrored on a microscopic level in the brain: here, too, connectedness ensures survival. Neurons that establish strong interconnections with their fellows—those participating in Attractors—make it through the winnowing phase. Those that do not join in stable bonds wither and drop from the consolidating template.

Consider how a baby learns to discern mere noise from the clicks, whistles, and burbles of speech. The human vocal cords, pharynx, tongue, lips, teeth, and palate can make thousands of distinct bits of sounds—*phonemes*—that meld seamlessly into words. An infant brain can hear and distinguish *all* possible phonemes; his genes carry the design for this mechanism. His panphonemic capacity prepares him to meet all human languages, but soon it becomes neural extravagance. Any language uses only a small subset of phonemes. English gets by with forty. Consequently, a toddler's brain contains Attractors for the phonemes that match his native tongue; only those linguistic sounds can he correctly hear and vocalize. Auditory experience has whittled his multipurpose scaffold into a purposeful template.

After the unused neurons expire, the truncated network no longer represents certain bits of knowledge. Japanese does not differentiate between the English sounds of "r" and "l," and a child steeped in Japanese hears no difference between them. French has no "th"—the Gallic world usually approximates the unpronounceable sound with "z"—stocking Pepé Le Pew's amorous entreaties with *zee, zis,* and *zat.* French poses similar sonorous barriers to English speakers: the gravelly, guttural "r" or the short, sharp "u" (not at all like the "oo" in *tube* or even the "yoo" in *unique*). And the vowel sound in the French word for "eye"—*oeil*—English has never heard its like. Only the exceptionally rare Anglo will pass Parisian muster on these pronunciation points.

This developmental progression—general scaffold, vigorous neuronal thinning, specialized template, and evolving configuration—unfolds in most neural systems, including limbic ones. An infant's emotional scaffold provides for temperament and for innate abilities like reading facial expressions. Limbic contact with his parents hones that pluripotential structure into the template of emotional life—the neural core of emotional identity. Once this quintessence is firm, we can say that a *person* exists, and we can know the individuated attributes of his emotional self. Ongoing experience gradually transforms his neural configuration, changing him from who he *was* into who he *is,* one synapse at a time. Emotional identity drifts over a lifetime—if fast and far enough, one may encounter a stranger's heart where a friend's or a lover's once dwelt.

Alfred de Musset, on seeing the novelist George Sand (the *nom de plume* of Amandine-Aurore-Lucile Dudevant), long after their love affair ended:

> *My heart, still full of her,*
> *Traveled over her face, and found her there no more . . .*
> *I thought to myself that a woman unknown*
> *Had adopted by chance that voice and those eyes*
> *And I let the chilly statue pass*
> *Looking at the skies*

IN THE BEGINNING

The first influence on the emotional mind is the origin of life itself: the double spirals that make up DNA. One filament, the *sense* strand, contains the linear sequence of instructions for building proteins. Its mirror image, the *nonsense* strand, encodes for nothing at all. Split the two, and fresh, complementary filaments assemble: across from sense the matching nonsense strand forms, and opposite nonsense, newly fashioned sense floats into existence. Even at this gritty biochemical level the duality of information, the practical and the poetic, lives. Half of the DNA doublet seems to signify nothing, but within its apparent biochemical inutility resides pure potential. The nonsense strand is always one cell division away from generating anew the pragmatic protein-building algorithm of its alter ego. This magical quality allows DNA to replicate, as it has for billions of years. And for the past hundred million years or so, when the echidna's ancestor diverged from the reptilian line, successful mammals have passed on the genes that built their limbic brains—to every one of us.

Genes can generate a limbic scaffold slanted toward shyness or a short temper as assuredly as they do long bones or a fair complexion. Dogs can be bred for their emotionality and usually are: breeders rely on the genetic transmission of temperament as they select for docility in the spaniel and ferocity in the pit bull. Certain strains of mice are thirty times more anxious than others; so are some human families. One brain's blueprint may promote joy more readily than most; in another, pessimism reigns. Whether happiness infuses or eludes a person depends, in part, on the DNA he has chanced to receive.

Is the gene train a juggernaut? Not at all. The limits to the genetic influence on personality are inscribed along the ovoid curve of the female pelvis. Over the last few million years, the primate brain has expanded faster than the bony outlet that is the baby's portal to an air-breathing world. If an infant is to squeeze out while his head still fits, his brain at birth can be only a fraction of its final size. He must defer most of his neural maturation until he leaves the womb—when his physiology is no longer flying solo but joins his parents' through their shared limbic nexus. His neurogenetic inheritance then becomes subject to the power of parental love.

CASTING THE MOLD

While genes are pivotal in establishing some aspects of emotionality, experience plays a central role in turning genes on and off. DNA is not the heart's destiny; the genetic lottery may determine the cards in your deck, but experience deals the hand you can play. Scientists have proven, for example, that good mothering can override a disadvantageous temperament. They arranged for especially nurturing monkey mothers to adopt baby monkeys genetically prone to anxiety. Anxious young monkeys usually become inhibited, low-ranking adults. The substitution of an attentive mother reversed their fates—once on a genetic path to a lifetime of timidity, these well-loved monkeys became dominant in their troops. The inverse also holds: inadequate nurturance can disrupt a healthy limbic inheritance, imposing anxiety and depression on someone who had the genetic makings of a happy life.

Like most of their toys, children arrive with considerable assembly required. A child's brain cannot develop normally without the coordinating influence that limbic communication furnishes. The coos and burbles that infants and parents exchange, the cuddling, rocking, and joyous peering into each other's faces look innocuous if not inane; one would not suspect a life-shaping process in the offing. But from their first encounter, parents guide the neurodevelopment of the baby they engage with. In his primal years, they mold a child's inherited emotional brain into the neural core of the self.

LEARNING TO SEE

In faith, I do not love thee with mine eyes,
For they in thee a thousand errors note,
But 'tis my heart that loves what they despise,
Who in despite of view is pleas'd to dote;
Nor are mine ears with thy tongue's tune delighted,
Nor tender feeling to base touches prone,
Nor taste, nor smell, desire to be invited
To any sensual feast with thee alone;
But my five wits, nor my five senses can

Dissuade one foolish heart from serving thee,
Who leaves unsway'd the likeness of a man,
Thy proud heart's slave and vassal wretch to be:
Only my plague thus far I count my gain,
That she that makes me sin awards me pain.

The list of the five senses cited in this sonnet is now passé. Based on an analysis of peripheral nerve endings, neuroscientists tabulate a longer list of independent senses: smell, sight, sound, taste, light touch, deep touch, the abilities to apprehend vibration, pain, and the position of one's joints. The poet who penned these lines did so in ignorance of this late-breaking news, but he knew something more important: emotionality is a different *kind* of sensory capacity, supplementary and integrative. The limbic brain relies on information supplied by the traditional inroads of perception, but it converts these data into a higher-order experience that goes beyond a catalogue of visual, auditory, or tactile qualities. The emotional whole—love's bottom line—is more than the sum of its sensory parts.

Consider the notorious Necker cube:

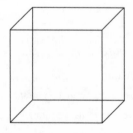

A two-dimensional bug crawling on the page could enumerate the essential elements of this figure: twelve straight lines, meeting at right, obtuse, and acute angles. True. But a bug from Flatland cannot see the scattered segments convening to depict the third dimension, rising out of the page and extending back into it. Because our brain is wired for 3-D, what we see extends beyond a dozen lines—so much that we cannot will ourselves to see only listless sticks. Now suppose the two-dimensional bug is a reptile, and let the lines represent the inner states of other organisms. Reptiles remain woodenly unresponsive to limbic dimension. Emotionality lifts our life experience out of reptilian Flatland; it makes the state of another creature's insides *matter.*

A child is born with the hardware for limbic sensing, but to use it skillfully he needs a guide. Someone must sharpen and calibrate his sonar; someone must teach him how to sense the emotional world correctly. Nor should this surprise us: experience is a necessary ingredient for normal sensory neurodevelopment. A child's brain will not germinate the neural machinery necessary for depth perception without input from both eyes. Limbic systems also need training on the right experiences to achieve full potential. These orienting experiences originate from an attuned adult. If a parent can sense her child well—if she can tune into his wordless inner states and know what he feels—then he, too, will become skilled in reading the emotional world.

A child makes constant use of his limbic link to adjust his impressions. The drama plays out a dozen times each summer afternoon at a local park. A toddler lurches across the grass with a determination that his unsteadiness renders positively quixotic. Inevitably gravity catches up with inexperience; he teeters and falls. At once he checks a parent's face: if she shows alarm or concern he cries, and if she is amused he may smile at her, even laugh. He trusts her assessment of his tumble more than his, and he does so with good reason. He can feel his pain and fright and disappointment but cannot gauge them. If his tumble is big enough to be awful or small enough to be negligible, he may realize that. But at all levels in between, he holds his emotions open to an expert's interpretation. A limbically attuned mother can tell a fearsome fall from a harmless one. When a child senses his mother's fear, his anxiety rises or falls in harmony with hers. He looks to his mother as a piano tuner looks to the sound of pure C. After he compares what he feels with what his mother shows, a child's emotional read on the world moves closer to hers.

Emotional experience begins as a derivative; a child gets his first taste of his feelings secondhand. Only through limbic resonance with another can he begin to apprehend his inner world. The first few years of resonance prepare this instrument for a lifetime's use. One of a parent's most important jobs is to remain in tune with her child, because she will focus the eyes he turns toward inner and outer worlds. He faithfully receives whatever deficiencies her own vision contains. A parent who is a poor resonator cannot impart clarity. Her inexactness smears his developing precision in reading the emotional world. If she does not or cannot teach him, in adulthood he will be unable to sense the inner states of others or himself. Deprived of the limbic compass that orients a person to his internal landscape, he will slip through his life without understanding it.

In Woody Allen's film *Deconstructing Harry,* an actor develops a sudden case of the blurs. At first his film crew thinks the lens is dirty, but they clean the camera and determine that the outline of the actor himself is smudged. "I don't know how to tell you this, but you . . . you're out of focus," a coworker tells the mortified player. "Mel—now, look—I want you to go home, and get some rest. See if you can just sharpen up," advises his director. At home, matters are not improved: "Daddy, you're all blurry!" says his dismayed child.

Allen's ability to concretize the abstractions of the human condition, rendering them simultaneously immediate and ludicrous, is central to his comedic gift. Fuzzy people exist, he tells us in this scene, people whose selves, not their bodies, are painfully indeterminate. Such a person enters psychotherapy because he does not know who he is. To people who *do* know, the predicament sounds improbable. But a person cannot know himself until another knows him. Omit skilled limbic resonance from the life of a child, and he will emerge with a psyche as indistinct as the blurry habitus of Allen's character. If a parent actively hates a child, if she affirmatively *knows* him in the punishing clarity of her fury—that child will fare better than one who languishes in the dim ether of emotional ignorance.

LEARNING TO BE

A baby begins life as an open loop. His mother's milk provides nourishment, and her limbic communication provides synchronization for his delicate neural rhythms. As a child matures, his neurophysiology internalizes some regulatory functions. Balanced from the outside in, his brain learns stability.

Lengthy parental absence deprives a child of limbic regulation. If he is very young, losing his parents upends his physiology. Prolonged separations even can be fatal to an immature nervous system, as vital rhythms of heart rate and respiration devolve into chaos. Sudden infant death is increased fourfold in the babies of mothers who are depressed—because without emotional shelter, infants die. The heart rhythms of securely attached babies are steadier than those with insecure relationships, just as the breathing teddy bear regularizes the respiration of premature infants. Synchronicity with parents (or, in a pinch, with another reliable rhythmic source) becomes the baby's developing physiologic strength.

Being well regulated in relatedness is the deeply gratifying state that people seek ceaselessly in romance, religions, and cults; in husbands and wives, pets, softball teams, bowling leagues, and a thousand other features of human life driven by the thirst for sustaining affiliations. In early life, limbic regulation is not simply pleasure; it is also crucial training. As a baby's precarious neurophysiology falls under the steadying spell of his mother, he first borrows her equipoise and then makes it his own. A child balances his physiology in the same way he masters a two-wheeler. A good parent rights him when he strays from the vertical; through repetition, a child absorbs the ability to correct his stance. Without words, concepts, ideas, or understanding, the vestibular and motor systems of his brain learn to do what the parent standing beside him once did.

Before a child can say "bicycle" (much less ride one), he is modulating his emotions via an external source. A distraught baby reaches for his mother because an attuned parent can soothe him; he cannot soothe himself. As a consequence of thousands of these interactions, a child learns to self-quiet. His knowledge, like knowing how to keep a bike upright, is implicit—invisible, inarticulate, undeniable.

The child of emotionally balanced parents will be resilient to life's minor shocks. Those who miss out on the practice find that in adulthood, their emotional footing pitches beneath them like the deck of a boat in rough waters. They are incomparably reactive to the loss of their anchoring attachments—without assistance, they are thrown back on threadbare resources. The end of a relationship is then not merely poignant but incapacitating.

An event as common as a cross-country move can uncover this sensitivity. In less time than it takes to read a good book, the average person can transport himself several thousand miles away from everyone and everything he has ever cared for. The limbic brain registers the disorienting loss of attachments as the all-purpose ache of homesickness. Letters and phone calls are a salve on the wound, but they are insubstantial substitutes for the full-bandwidth sensory experience of nearness to the ones you love. To sustain a living relationship, limbic regulation demands sensory inputs that are rich, vivid, and frequent.

Most therapists have in their practices at least one casualty of a calamitous move—the college student leaving home for the first time, the recipient of a distancing promotion or transfer—whose seeming psychological health collapses under the geographic strain. Often no one is more surprised than the sufferer, who had no idea of his emotional frailty and the support networks of his native environs.

Our society overlooks the drain on emotional balance that results from severing attachments. From the dawn of the species until a few hundred years ago, most human beings lived out their lives in one community. The signature lesson of the twentieth century is that unforeseen complications are ever the faithful companions of technological progress. The convenient devices that enable extensive mobility are problematic because limbic regulation operates weakly at a distance. We have the means to establish a peripatetic lifestyle, but we will never have the brains for it.

LEARNING TO LOVE

Suppose for a moment that a friend whisks you to New York's Museum of Modern Art. The dusty basement of a château undergoing renovation has yielded up a new painting, he tells you, and it is on display. The painter was either Manet or Monet, or perhaps Matisse—your friend cannot recall, his mind awhirl in French *artistes* and their alliterative surnomic M's. The painting hangs unlabeled. If you are familiar with the works of the three artists, you may be confident that you could tell them apart.

How? By the reflected gleam of implicit memory. If you have seen, say, Matisse's work, then you hold title to an accessible prototype that imparts what his paintings generally look like. A critic or connoisseur might be able to articulate the identifying characteristics: composition, lighting, tints, texture, perspective, subject, color. But book-learning is superfluous here. Armed only with the neural results of serial viewing, absorbing, pondering—some paintings *look* and *feel* like Matisse. Others do not. Without a degree in art history, you can know a Matisse when you see one. As you see more of them your intuition sharpens.

So it is with emotional knowledge. In the first years of life, as his brain passes from the generous scaffold to the narrow template, a child extracts patterns from his relationships. Before any glimmerings of event memory appear, he stores an impression of what love *feels* like. Neural memory compresses these qualities into a few powerful Attractors—any single instance a featherweight, but accumulated experience leaves a dense imprint. That concentrated knowledge whispers to a child from beneath the veil of consciousness, telling him what relationships *are*, how they function, what to anticipate, how to conduct them. If a parent loves him in the healthiest way, wherein his needs are paramount, mistakes are forgiven, patience is plentiful, and hurts are soothed as best they can be, then *that* is how he will relate to himself and others. Anomalous love—one where his needs don't matter, or where love is suffocating or autonomy intolerable—makes its ineradicable limbic stamp. Healthy loving then becomes incomprehensible.

Zeroing in on *how* to love goes hand in hand with *whom*. A baby strives to tune in to his parents, but he cannot judge their goodness. He attaches to whoever is there, with the unconditional fixity we profess to require of later attachments: for better or worse; for richer, for poorer; in sickness and in health. Attachment is not a critic: a child adores his mother's face, and he runs to her whether she is pretty or plain. And he prefers the emotional patterns of the family he knows, regardless of its objective merits. As an adult his heart will lean toward these outlines. The closer a potential mate matches his prototypes, the more enticed and entranced he will be—the more he will feel that here, at last, with this person, he *belongs*.

It is attachment that makes familiarity trump worth. A golden retriever thrills only to *his* owner. He is amiably and helplessly indifferent to passersby who may be kinder, fonder of walks, quicker with treats—he does not, he *cannot* value them. Everyone is in the same limbic boat as those patient, expectant dogs.

Most people accept without difficulty the neural miracles of exactitude contained in 20/20 vision or perfect pitch. But some balk at believing that a person can scan a crowd and pick out the intimate elements in a stranger's heart. Can somebody survey a group and intuit who has a bad temper, an alcoholic mother, who dreams at night of revenge on the father who left him? Look at the relationships around you and judge for yourself. People target the mates who mesh with their own minds, and they do so with speed and precision that our smartest smart bombs are not sufficiently intelligent to envy.

A relationship that strays from one's prototype is limbically equivalent to isolation. Loneliness outweighs most pain. These two facts collude to produce one of love's common and initially baffling quirks: most people will choose misery with a partner their limbic brain recognizes over the stagnant pleasure of a "nice" relationship with someone their attachment mechanisms cannot detect. Consider the young man described in the last chapter wrestling with the present-day reenactment of the long-ago love with his fiery, critical mother. As an adult, he faces a binary universe. If he connects with a woman, she turns out to be his mother's younger clone. But a supportive woman leaves him exasperatingly empty of feeling—no spark, no chemistry, no fireworks.

Many people can relate the love story they *think* they have in mind: boy meets girl (and every permutation thereof), they fall in love, and live happily ever after. But this story dwells in the airy regions of the cortex, which drafts its scripts using imagination, logic, and will. In the older, deeper, and occasionally darker structures of the limbic brain, a different trio cooperates: attachment, implicit memory, and strong Attractors. There one can read love stories like this: boy meets girl, who (reminiscent of his mother) is needy and stifles his independence; they struggle bitterly over the years and resent each other a little more every day. Some people carry that tale in their hearts, and whether they find a player for the part or not, the piece can only come to grief.

The limbic Attractors that form in childhood can be multiple. A single relationship or household can spawn as many Attractors as it embodies predictable lessons. Thus a child can form influential Attractors from relationships not just with mother and father, but also with siblings, nannies, even the family as a whole. In a home with ten children, for instance, each may extract a version of the local truth that there isn't enough love to go around in the world, that you must fight fiercely and ceaselessly and still your heart will go hungry. A newspaper column with advice for parents recently advocated letting older and younger children settle disputes without parental interference, so that they might learn what people do in the real world. The children in the unfortunate households where parents apply this pearl will unerringly distill the timeless lesson of the unsupervised boarding school or playground: justice is weak; might and intimidation triumph.

The consequences that flow from early limbic lessons entail a final complexity: emotional reality (or its illusion) is collaborative. Back at the MOMA, you inspect the painting that looks like a Matisse. Your formerly forgetful friend snaps his fingers and says he remembers that he heard the artist is Manet. As he stands next to you, radiating Manet certainty, the painting itself starts to change before your eyes. Pigments fade and oxidize, lines blur and reform, and the painting, if not a textbook Manet, is now indisputably *more* Manet than before.

Visual virtuality is mildly susceptible to interference from others, and limbic virtuality is a good deal more so. As every child knows, if one passes a magnet over a handful of sand, thousands of tiny magnetically sensitive particles—iron filings—leap to the magnet, while the silicate crystals of stone remain unmoved. The limbic brain is an emotional magnet. Attractors activate compatible aspects of relatedness and emotionality in others, leaving dormant the incompatible pebbles. We all embody an emotional force field that acts on the people we love, evoking the relationship attributes we know best. Our minds are in turn pulled by the emotional magnets of those close to us, transforming any landscape we happen to contemplate and painting it with the colors and textures *they* see.

The young man with a fondness for faultfinding lovers is in even more trouble than he thinks. First, he must contend with the mental mechanism that leads him with uncanny precision to a woman who is *herself* critical. Second, *his* presence will magnify whatever minatory tendencies his current paramour may possess. Ditto for her: she has chosen her man because he matches an Attractor of hers, and she will enhance the matching virtues and vices.

EVER AFTER . . .

After the formative years that instill templates and Attractors, emotional learning doesn't stop, but it slows. Childhood chisels its patterns into pliable neural networks, while later experience wields a weaker influence on the evolving person. Why should this be? In theory, the same learning so influential in casting the emotional core of the self could take place later in life. But often the only emotional learning one sees after childhood is the reinforcement of existing fundamentals.

Unfortunately, the brain's biology and its mathematics both oppose adult emotional learning. The plasticity of the brain—the readiness of neurons to sprout fresh connections and encode new knowledge—declines after adolescence. And later learning is energetically unfavorable within a neural network. New lessons must fight an uphill battle against the patterns already ingrained, because existing Attractors can easily overwhelm and absorb moderately novel configurations. The nature of neurovirtuality ensures that it trims the ambiguity from reality, and portrays largely what has already been seen. And so, left to his own devices, a child who knew and loved a deceitful, selfish, or jealous parent does not often learn to love differently at age twenty, forty, or sixty.

Every child stores his Attractors and lives out his adult life on a neurally generated soundstage. If his Attractors mislead him, is there any way out? Can he manage to escape the trap of seeing what he's always seen and doing what he's always done? Once emotional learning has gone awry, can it be set right?

Despite the longevity of Attractors and the waning of neural flexibility, the emotional mind *can* change in adulthood. The old patterns can undergo revision, although the task is not easy. You may think that we have given late emotional learning short shrift in this brief section, but the topic is so near to our hearts that it merits a chapter to itself.

Eight

BETWEEN STONE AND SKY

WHAT CAN BE DONE TO HEAL HEARTS GONE ASTRAY

The real voyage of discovery consists not in seeking new landscapes, but in having new eyes.

—Marcel Proust

In the heart of Rome stands the world's most magnificent fountain: Trevi, named for its location at the juncture of three roads (*tre vie*). Poseidon rises tall in his chariot as stamping horses pull him through an angry sea; sheets of water cascade down rocks to the basin below. As legend has it, the traveler who tosses a coin into the fountain's pool will ensure his return to the Eternal City.

Imagine that between one breath and the next, time freezes. The glittering coruscations at Trevi hang immobile. One high-flung drop floats in the air, a gleaming jewel against the motionless blue of the Italian sky.

We might struggle to determine the history of the aqueous pearl—how the meaningless plungings of water and wind produced *that* globular permutation of molecules. For any feature of the physical world, drawing on knowledge of the relevant forces, we can look down the corridors of time and frame a sequence that could have culminated in the outcome before us. The question a developmentalist seeks to answer, whether of a raindrop or an emotional mind: *how did it turn out this way?* Starting with the same structure, and pivoting on time's axis to face the future, a different riddle confronts: *what is the fate of this system?* What contour will it assume after a minute, an hour, a year?

The side street of mathematics called *chaos theory* declares such a pursuit impossible. With time arrested, the location of those teardrop molecules is knowable. Unstop the tableau, allow a moment to pass, and the drop disperses. Nobody can foresee where its component molecules will end up a few seconds later. In the city of Rome, yes; in the fountain of Trevi, probably, but *exactly* where? The universe denies a more specific glimpse to any being less than God. The surging possibilities attending turbulence are too numerous for mortals to tabulate.

So it is with the emotional mind. If the fountain water were frozen as solid as the rock that cups it, we could predict its conformation over the next minute with certainty. If *all* possible arrangements of water molecules were equally probable, we could throw up our hands in graceful defeat before infinity. It is the *liquidity* of the water and the mind that befuddles, their ability to assume an array of forms with immense—but not limitless—variety. Like a bead of sea spray, the future of an emotional mind hangs between the immobility of stone and the freedom of the summer sky. Identity can change, but only within the outlines its architecture commands.

So much depends on the emotional learning that adult neurophysiology permits. Can the neglected or abused child hope for a healthy life? Will his adulthood replicate his past and prove again the principles he knows too well? Considering the neural impediments to progress, how does healing happen? With Attractors ready to shoehorn reality into the mold of the familiar, how does an emotional mind break free?

Psychotherapy grapples with these questions daily. A therapist does not wish merely to discern the trajectory of an emotional life but to determine it. Helping someone escape from a restrictive virtuality means reshaping the bars and walls of a prison into a home where love can bloom and life flourish. In the service of this goal, two people come together to change one of them into somebody else.

Few agree on *how* the metamorphosis occurs. The secret identity of psychotherapy's mutative mechanism has prompted enough hot-tempered debate and factional feuding to fill a history of the Balkans. And rightly so. The centerpiece of therapy is also the focal point of the human heart.

PARADIGMS LOST

Twin factions warred over the mind up and down the length of the twentieth century. Smoke and clamor from the battle still obscure psychotherapy's central concern.

On one hand, a "biological" group holds that mental events emanate from the material world of the brain. Mental pathology therefore begins in deformations of that physicality: misshapen receptors, defective genes, brain damage. All true. This school favors remedies like medications, electric current, and magnetic fields. Sometimes they work.

On the other hand, the venerable "psychological" cadre sees emotional disturbances flowing from an intangible realm, wherein memory's ghosts walk, feelings have force, and relationships order themselves on past patterns. Again true. This camp advocates cure through a somewhat bewildering multiplicity of psychotherapies—also intermittently effective.

Each party to the fracas cites evidence favorable to its side, as if the weight of unilateral testimony could tip the scales to victory. But dividing the mind into "biological" and "psychological" is as fallacious as classifying light as a particle or a wave. The natural world makes no promise to align itself with preconceptions that humans find parsimonious or convenient. As it happens, light confounds the deceptively simple dichotomy that beckoned to scientists for decades. Every experimenter who tried to prove light particulate succeeded, as did every test of its wave nature. Impossible, in theory. Particle and wave are mutually negating ideas; a thing cannot be both itself and its opposite. In reality, "particles" and "waves" exist in minds, not in nature. These crude categories cannot capture the essence of light.

The emotional mind likewise transcends the facile and appealing dualism separating its psychological and biological aspects. Physical mechanisms produce one's experience of the world. Experience, in turn, remodels the neurons whose chemoelectric messages create consciousness. Selecting one strand of that eternal braid and assigning it primacy is the height of capriciousness. In our post-Prozac nation, most are aware that modern medications can modify personality traits. Of less renown is the reciprocal finding (provided by advanced imaging technologies) that psychotherapy alters the living brain. The war over the mind can be halted and a truce proclaimed, but only because both armies have always occupied all the territory. As the Dodo observes to Alice in Wonderland, everybody has won, and *all* must have prizes.

The mind-body clash has disguised the truth that psychotherapy *is* physiology. When a person starts therapy, he isn't beginning a pale conversation; he is stepping into a somatic state of relatedness. Evolution has sculpted mammals into their present form: they become attuned to one another's evocative signals and alter the structure of one another's nervous systems. Psychotherapy's transformative power comes from engaging and directing these ancient mechanisms. Therapy is a living embodiment of limbic processes as corporeal as digestion or respiration. Without the physiologic unity limbic operations provide, therapy would indeed be the vapid banter some people suppose it to be.

"Where id was, there ego shall be" was Freud's battle cry, a magisterial encapsulation of the talking cure as prolonged explanation. Freud saw insight and intellect vanquishing the mind's dark undergrowth like conquistadors beating back jungle to build a city.

Speech is a fancy neocortical skill, but therapy belongs to the older realm of the emotional mind, the limbic brain. Therapy should not seek to overrule the primeval forces predating civilization, because, like love, therapy is already one of them.

People do come to therapy unable to love and leave with that skill restored. But love is not only an end for therapy; it is also the means whereby every end is reached. In this chapter we will examine how love's three neural faces—limbic resonance, regulation, and revision—constitute psychotherapy's core and the motive force behind the adult mind's capacity for growth.

CHANGING THE EMOTIONAL MIND

LIMBIC RESONANCE

Every person broadcasts information about his inner world. As a collection of dense matter betrays its presence through electromagnetic emissions, a person's emotional Attractors manifest themselves in a radiant aura of limbic tones. If a listener quiets his neocortical chatter and allows limbic sensing to range free, melodies begin to penetrate the static of anonymity. Individual tales of reactions, hopes, expectations, and dreams resolve into themes. Stories about lovers, teachers, friends, and pets echo back and forth and coalesce into a handful of motifs. As the listener's resonance grows, he will catch sight of what the other sees inside that personal world, start to sense what it feels like to live there.

Therapists are sometimes tempted to catalogue and analyze the output of a patient's volubility—an inviting but hollow detour. Take a few measures from the Italian composer Ottorino Respighi's "Fountains of Rome," a tone poem meant to evoke (among others) Trevi. How can its meaning be disclosed? One could dissect the notes, scrutinize the sound frequencies, chart and measure the silent intervals. But anyone wishing to receive what

Respighi has to say need only listen. Part of the brain enables us to assemble certain sounds in a loftier coherent dimension. As a result, Respighi's exuberant outpouring requires no schooling to grasp. Music, said Beethoven, is a higher revelation than philosophy. Another part of the brain is poised to translate emotional signals into revelations higher still. This music a therapist ignores at his peril.

The first part of emotional healing is being limbically known—having someone with a keen ear catch your melodic essence. A child with emotionally hazy parents finds trying to know himself like wandering around a museum in the dark: almost anything could exist within its walls. He cannot ever be sure of what he senses. For adults, a precise seer's light can still split the night, illuminate treasures long thought lost, and dissolve many fearsome figures into shadows and dust. Those who succeed in revealing themselves to another find the dimness receding from their own visions of self. Like people awakening from a dream, they slough off the accumulated, ill-fitting trappings of unsuitable lives. Then the mutual fund manager may become a sculptor, or vice versa; some friendships lapse into dilapidated irrelevance as new ones deepen; the city dweller moves to the country, where he feels finally at home. As limbic clarity emerges, a life takes form.

Limbic Regulation

BALANCE THROUGH RELATEDNESS

Certain bodily rhythms fall into synchrony with the ebb and flow of day and night. These rhythms are termed *circadian*, from the Latin for "about a day." A more fitting appellation is *circumlucent*, because they revolve around light as surely as Earth. Human physiology finds a hub not only in light, but also in the harmonizing activity of nearby limbic brains. Our neural architecture places relationships at the crux of our lives, where, blazing and warm, they have the power to *stabilize.* When people are hurting and out of balance, they turn to regulating affiliations: groups, clubs, pets, marriages, friendships, masseuses, chiropractors, the Internet. All carry at least the potential for emotional connection. Together, those bonds do more good than all the psychotherapists on the planet.

Some therapists recoil from the pivotal power of relatedness. They have been told to deliver insight—a job description evocative of estate planning or financial consulting, the calm dispensation of tidy data packets from the other side of an imposing desk. A therapist who fears dependence will tell his patient, sometimes openly, that the urge to rely is pathologic. In doing so he denigrates a cardinal tool. A parent who rejects a child's desire to depend raises a fragile person. Those children, grown to adulthood, are frequently among those who come for help. Shall we tell them again that no one can find an arm to lean on, that each alone must work to ease a private sorrow? Then we shall repeat an experiment already conducted; many know its result only too well. If patient and therapist are to proceed together down a curative path, they must allow limbic regulation and its companion moon, dependence, to make their revolutionary magic.

Many therapists believe that reliance fosters a detrimental dependency. Instead, they say, patients should be directed to "do it for themselves"—as if they possess everything but the wit to throw that switch and get on with their lives. But people do not learn emotional modulation as they do geometry or the names of state capitals. They absorb the skill from living in the presence of an adept external modulator, and they learn it implicitly. Knowledge leaps the gap from one mind to the other, but the learner does not experience the transferred information as an explicit strategy. Instead, a spontaneous capacity germinates and becomes a natural part of the self, like knowing how to ride a bike or tie one's shoes. The effortful beginnings fade and disappear from memory.

People who need regulation often leave therapy sessions feeling calmer, stronger, safer, more able to handle the world. Often they don't know why. Nothing obviously helpful happened—telling a stranger about your pain sounds nothing like a certain recipe for relief. And the feeling inevitably dwindles, sometimes within minutes, taking the warmth and security with it. But the longer a patient depends, the more his stability swells, expanding infinitesimally with every session as length is added to a woven cloth with each pass of the shuttle, each contraction of the loom. And after he weaves enough of it, the day comes when the patient will unfurl his independence like a pair of spread wings. Free at last, he catches a wind and rides into other lands.

BALANCE THROUGH MEDICATION

A limbic connection can steady a person whose emotions are tumbling out of control. But some states are beyond the power of attachment to modulate. Limbic regulation does relatively little to remedy problematic adult temperaments, for instance. Severe depression is also outside the reach of relationships. Depression often leads a person to shun social contact, nullifying the regulatory impact of his affiliative ties. Even when he does interact, a depressed person is likely to avert his gaze, cutting himself off from the interchange of emotional signals. And depression shuts down limbic circuits: in one study, depressed patients were no more able to recognize facial expressions than patients who had sustained brain damage to the area responsible for that function. Thus does depression render someone immune to the healing force in others that might counterbalance his despair.

Medication can sometimes steer emotions when attachment cannot. Directly manipulating the neurochemistry of emotion is a tricky enterprise—exciting in its promise, and frightening in the scope of the damage if the intervention is inexpert. Using medications to alter the emotional mind means tinkering with the stuff the self is made of. In the right hands, that alchemy can rescue lost lives.

A psychiatrist who employs medication can count on meeting some stiff opposition. Treating major depression with chemicals invites a paradigm clash between physician and patient rivaling the collision between Galileo and the Pope. One side recounts with tenacity the iron inevitability of despair, durable hopelessness, monolithic bleakness, pain, apprehension, horror, and death. The other side answers with the prospect of optimism in pill form.

From the patient's perspective, the physician's claim is other-worldly. Depression's dark prism has often thrown into shadow whatever credibility he might have extended to medication in his happier days. Now, evaluating all propositions is an uncertain business. Like any other momentous shift in emotion, depression is not an occupation by a foreign army; it is civil insurrection, the subversion of identity's republic from within. A depressed person loses more than energy and appetite—he loses himself and the ca-pacity to make the decisions his former, pre*coup* self would have made.

If he takes an antidepressant, therefore, it is rarely for logic's sake. He can't see and does not wholly (or sometimes remotely) be-lieve in the sunnier world the psychiatrist dwells in. The oblong capsule becomes for him a crucifix, a Star of David, a cross of Lor-raine: the emblem of faith in the promise of a better world.

The seed of that trust must precede psychotherapy and psy-chopharmacology both. Psychiatrists seldom advertise the prereq-uisites for treatment. Perspicacity is optional; a patient doesn't have to spy any reality but his own. Indeed, he usually cannot—it's a therapist's job to span two worlds at once. But a patient has to stomach the proposition that his emotional convictions are fiction, and someone else's might be better. Not everyone can do it. A psychiatrist's office should bear a placard analogous to the posted minimum height for roller coasters: YOU MUST BE AT LEAST THIS TRUSTING TO RIDE THIS RIDE.

One young woman, for instance, demanded that her therapist explicate a supporting framework to bolster what she saw as an absurdly slender claim of trustworthiness. "Why should I believe you, and not myself?" she asked doggedly. "Give me one good reason." Hers seemed a sound request; the two searched and argued and pondered for months. They found no reason because none exists. A psychiatrist's training and education, his credentials, his years of practice, establish nothing absolutely. An authority can be wrong, and a novice correct (if by accident), on any issue. A seasoned professional, while more *likely* to be right on topics falling within his domain, can neither prove nor guarantee his rectitude where two virtualities meet. Psychiatry runs on the same elixir that fuels the rest of medicine: a fervent wish that somebody else knows better. People who trust a little can gamble and learn to trust more; people who have no faith from which to leap are out of luck. Mental health is a substance that attracts itself as readily as money or power: the more you have, the more you can get.

Some find trust an incongruous companion in the pharmacological treatment of emotional illness. After all, shouldn't the potency and reach of modern medications effectively supplant reliance on ancient faith-healing tools? Since the patients in question are inescapably social creatures, the answer is no. Placebo response rates for depressive conditions routinely scatter in the 30 percent range; for anxiety disorders, 40 percent and up. Some observers have erroneously taken this data to mean that a placebo does nothing and that if drugs can't do better, they must be exceedingly weak. Quite the opposite. The result of the limbic interaction between patient and healer is so efficacious that only the most powerful medications can be definitively *proven* stronger.

For the depressed person, medication can be an ax for the frozen sea within. Sleep and appetite respond first, gliding back into alignment from their former deviations. Spouses and friends begin to catch inklings of someone fractionally more familiar. Interest returns, then pleasure, and finally, the ability to laugh. Bleak ideas recede in relevance; morbid thoughts slip away down dim corridors. After several months, depression may fade as a bad dream does by midmorning, leaving only an afterimage of unpleasantness.

Given the dual agents in his armamentarium—one of them human, one chemical, both limbically formidable—a psychiatrist must decide when to use each. The first issue is easy: the best of biologically minded healers always wield the limbic component, since it is inherent, effective, and side-effect free. But when to prescribe an agent other than the doctor himself? In some cases, medications are literally lifesaving: major depression and bipolar disorder still claim thousands each year. Easing frank suffering and curbing emotional morbidity are goals that, in these pharmacologically sophisticated times, most people accept without question.

Still, many clinical encounters do not contain the defining element of such urgency. Often, patients contemplate taking medication not to stave off death or a recognized illness, but simply to help them reach a happier state. To some people, a pharmacological route smacks of an immoral shortcut, as though they are snatching a boon without enduring arduous qualifying trials that a stern universe surely demands. If a person enters psychotherapy and emerges, years later, with his moodiness banished or anxiety erased, nobody thinks he has cheated destiny or disapproving gods. If he takes a pill, and reaches the same goal in a few days or weeks, many will wonder—is that permissible, legitimate, *fair*? "People must *work* to better themselves" is an intuitive (and often correct) philosophy, one that in this instance finds a voice in the therapists, still numerous, who tell patients that medication and psychotherapy's edifying labor don't mix.

In reality, the naysayers have little to worry about. For most of history, humanity has employed a handful of emotional regulators—alcohol, opium, cocaine, cannabis, a few others. All have had major drawbacks. The truly effective chemical modulation of emotionality is a dazzling scientific achievement, even if the underlying mechanisms remain impenetrable mysteries. But medications cannot resolve all limbic predicaments, not by half. What they lack in nuance they make up in strength, but sometimes nuance is called for. Early emotional experiences knit long-lasting patterns into the very fabric of the brain's neural networks. Changing that matrix calls for a different kind of medicine altogether.

Limbic Revision

Knowing someone is the first goal of therapy. Modulating emotionality—whether by relatedness or psychopharmacology or both—is the second. Therapy's last and most ambitious aim is revising the neural code that directs an emotional life. Somewhere within a person's brain lie the myriad connections embodying his limbic knowledge—the strong Attractors that bend emotional perceptions and guide actions in love. When a therapist wants to help a patient who suffers from unfulfilling relationships or an immobilizing deficit in self-esteem, he wants to alter the microanatomy of another person's brain.

If any agency can build or destroy the bridges between neurons, strengthen or weaken them, then neural knowledge can change. But the brain has multiple learning systems, and all information does not change in the same way. Seven plus three equals ten, wrote Augustine, not now but always. "In no circumstances have seven and three ever made anything else than ten, and they never will. So I maintain that the unchanging science of number is common to me and to every reasoning being." Suppose Augustine spun in his grave, and the rules of mathematics underwent a convulsive shift that sent seven plus three hurtling all the way to eleven. Anybody could read this update in the morning newspaper and modify additions immediately. The neocortical brain collects facts quickly. The limbic brain does not. Emotional impressions shrug off insight but yield to a different persuasion: the force of another person's Attractors reaching through the doorway of a limbic connection. Psychotherapy changes people because one mammal can restructure the limbic brain of another.

REVISING RELATIONSHIP PATTERNS

A person cannot choose to desire a certain kind of relationship, any more than he can will himself to ride a unicycle, play *The Goldberg Variations,* or speak Swahili. The requisite neural framework for performing these activities does not coalesce on command. A vigorous self-help movement has championed the hoax that a strong-willed person, outfitted with the proper directions, can select good relationships. Those seduced into the promise of a quick fix gobble it up. But the physiology of emotional life cannot be dispelled with a few words. Describing good relatedness to someone, no matter how precisely or how often, does not *inscribe* it into the neural networks that inspire love.

Self-help books are like car repair manuals: you can read them all day, but doing so doesn't *fix* a thing. Working on a car means rolling up your sleeves and getting under the hood, and you have to be willing to get dirt on your hands and grease beneath your fingernails. Overhauling emotional knowledge is no spectator sport; it demands the messy experience of yanking and tinkering that comes from a limbic bond. If someone's relationships today bear a troubled imprint, they do so because an influential relationship left its mark on a child's mind. When a limbic connection has established a neural pattern, it takes a limbic connection to revise it.

An attuned therapist feels the lure of a patient's limbic Attractors. He doesn't just hear about an emotional life—the two of them *live* it. The gravitational tug of this patient's emotional world draws him away from his own, just as it should. A determined therapist does not strive to have a good relationship with his patient—it can't be done. If a patient's emotional mind would support good relationships, he or she would be out having them. Instead a therapist loosens his grip on his own world and drifts, eyes open, into whatever relationship the patient has in mind—even a connection so dark that it touches the worst in him. He has no alternative. When he stays outside the other's world, he cannot affect it; when he steps within its range, he feels the force of alien Attractors. He takes up temporary residence in another's world not just to observe but to alter, and in the end, to overthrow. Through the intimacy a limbic exchange affords, therapy becomes the ultimate inside job.

Each emotional mind formed within the force field of parental and familial Attractors. Every mind operates according to the primordial principles absorbed from that charged environment. A patient's Attractors equip him with the intuition that relationships feel like *this,* follow *this* outline. In the duet between minds, each has its own harmonies and the tendency to draw others into a compatible key. And so the dance between therapist and patient cannot trace the same path that the latter expects, because his partner moves to a different melody. Coming close to a patient's limbic world evokes genuine emotional responses in the therapist—he finds parts of himself stirring in response to the particular magnetism of the emotional mind across from him. His mission is neither to deny those responses in himself, nor to let them run their course. He waits for the moment to move the relationship in a different direction.

And then he does it again, ten thousand times more. Progress in therapy is iterative. Each successive push moves the patient's virtuality a tiny bit further from native Attractors, and closer to those of his therapist. The patient encodes new neural patterns over their myriad interactions. These novel pathways have the initial fragility of spring grass, but they take deep root within an environment that provides simple sustaining limbic nutrients. With enough repetition, the fledgling circuits consolidate into novel Attractors. When that happens, identity has changed. The patient is no longer the person he was.

Therapy's transmutation consists not in elevating proper Reason over purblind Passion, but in replacing silent, unworkable intuitions with functional ones. Patients are often hungry for *explanations*, because they are used to thinking that neocortical contraptions like explication will help them. But insight is the popcorn of therapy. Where patient and therapist *go* together, the irreducible totality of their mutual journey, is the movie.

Recall the child from chapter 7 who grows up hearing only Japanese. While English has two different phonemic Attractors for the discrete sounds of "r" and "l," Japanese has only a single broad Attractor encoding for an intermediate noise. A Japanese adult, armed with his original overlapping Attractor, hears no difference between "right" and "light."

Researchers recently trained Japanese adults to make the sonorous distinction for which their youth ill prepares them. Dr. Jay McClelland at the Center for the Neural Basis of Cognition played standard English conversations for Japanese speakers, and found that listening to common talk actually degraded any ability they had to discern "r" from "l." This result reflects an Attractor's *modus operandi:* each individual "r" and "l" fell into the broad basin of the Attractor's r-l funnel. The outer reality of two discrete intonations was reduced to the inner virtuality of one, and the mind's ear registered "rl." Reiteration naturally strengthened that unselective Attractor.

But when McClelland gave his subjects serial exposure to a purified "r," and "l," with accentuation of the formative sound characteristics, each person's inclusive r-l Attractor gradually divided into a pair of distinctive ones. Japanese adults could then distinguish "right" from "light" as well as any resident of Brooklyn or Manhattan Beach. McClelland's work demonstrates not only that the adult brain retains sufficient plasticity to encode fresh Attractors, but also that a specialized experiential environment can instill neural lessons when ordinary life cannot. Psychotherapy performs the same process on emotional discriminations.

The set of all possible relationship stories, all styles of loving that lead to misery, is illimitable. That infinitude makes the daily practice of psychotherapy a mind-expanding enterprise. When a patient first walks through the door, we can reasonably expect him to strike up the relationship he knows, but we have never seen one quite like it before. But a therapist doesn't need an encyclopedic compendium of every unhappy relationship variant. Instead, his indispensable tools are the strong template of healthy relatedness within himself, and the keen sense of wrong when he and the patient depart familiar territory.

When therapy modifies how someone lives a relationship, it corrects *whom* he may join in love. Decision cannot effect such an alteration. Knowing that a recurrent partner haunts you doesn't adjust a heart's direction. Many people suppose that therapy gives people a clear picture of a tormenting *amour* so they can spot and thereby avoid future deadly incarnations. Not so. You can't tell someone with faulty Attractors to go out and find a loving partner—from his point of view, there are none. Those who could love him well are invisible. Even if the clouds parted and a perfectly compassionate and understanding lover descended from heaven on a sunbeam to land at his feet, *his* mind would still be tuned to another sort of relationship; *he* still wouldn't know what to do. A wise therapist, paraphrasing T. S. Eliot, would advise him to wait without hope, because his hope would be hope for the wrong thing, and to wait without love, because his love would be love of the wrong thing.

Therapy doesn't clarify the object of desire so an intoxicated traveler can spend the rest of his life dodging it. Therapy worthy of the name changes what he wants. When he finishes, his heart tends in a healthier direction, the allure of former pathology diminishes, and what once was barely noticeable becomes his new longing.

If psychotherapy exerts its healing touch through limbic connections, one wonders, why aren't *other* attachments curative? If he were willing to put in the time—why couldn't a spouse, friend, bartender, or bowling partner guide a lost soul into a healthier emotional world?

The matter is one of probability rather than destiny. A person who needs limbic revision possesses pathologic Attractors. Everyone who comes within range feels at least some of the unhappiness inherent in his world, and that intimation repels many potentially healthy partners. Those who stay often do so because they recognize a pattern from their own pasts. For them it is a siren song. Relatedness engenders a brand loyalty that beer companies would kill for: your own relationship style entices. Others are wearisome and, in short order, unpalatable. Thus people who bond share unspoken assumptions about how love works, and if the Attractors underlying those premises need changing, they are frequently the last people in the world who can help each other.

And yet, on a planet of six billion personalities colliding and meeting with the frenetic energy of infinitesimal molecules in their perpetual Brownian dance, the improbable is occasionally bound to occur. A person with maladaptive Attractors *can* encounter another by chance who will teach him what he needs to learn. The instructor fate provides, whether husband or wife, brother, sister, or friend, is often amiably unmoved by the other's problematic emotional messages. Through the reach of their relationship and the utility of his relative imperviousness, he can gently and incrementally dissuade his student from headlong flight down paths that terminate in sorrow. Because of the tremendous variability in the configuration of human hearts and the randomness that throws people together, such felicitous combinations are as inevitable as they are precious. Against the odds, as it has since the beginning, life finds a way.

WHEN PSYCHOTHERAPY GOES ASTRAY

When a therapist establishes a limbic conduit to influence his patient, he simultaneously opens himself to the other's emotional Attractors. A therapist's odd gift lies in tuning into strange melodies enough to hear them, while he resists falling into complete harmony. This arrangement is plainly precarious, the gaping voids on either side of the tightrope all too visible. When therapy falters or fails (and that is far from rare), the twofold reasons are just what one might suppose: the mishap of missing the patient's limbic communication entirely, and the blunder of being swept into unpleasant alignment with foreign Attractors.

SOLID ICE

We should always remember that the work of art is invariably the creation of a new world, so that the first thing we should do is to study that new world as closely as possible, approaching it as something brand new, having no obvious connection with the worlds we already know. When this new world has been closely studied, then and only then let us examine its links with other worlds, other branches of knowledge.

Nabokov was setting forth the requirements for reading a novel, but he might as well have been describing the outlook most congenial to apprehending the parallel limbic realities of the people around us. A capable therapist shares much with a good reader: he must willingly suspend his belief in the rules he knows and approach a personal universe whose workings should be unimaginable to the uninitiated. If he is able to attain a state of sufficient receptivity, a therapist can allow the other mind to burst onto the scene like great art does—"as a more or less shocking surprise."

The therapist who cannot engage in this open adventure of exploration will fail to grasp the other's essence. His every preconception about how a person *should* feel risks misleading him as to how that person *does* feel. When he stops sensing with his limbic brain, a therapist is fatally apt to substitute inference for resonance.

Therapists prone to surrender limbic vision come from schools that offer cookie-cutter solutions. The formulaic assumptions revolve with the passing of years, but the mistake remains the same. At the turn of the century, emotional troubles were all due to penis envy and castration anxiety. The prevailing political climate forbids these ailments, but repressed memories and attention deficit disorder have taken their place as today's sacred afflictions. Tomorrow it will be something else.

How do some illnesses disappear, while others arise *de novo* with a flourish of psychopathological vitalism? Popular prejudices alternately obscure and exaggerate the prevalence of emotional ailments. But those seeking treatment have enough to worry about without being saddled with predetermined pathology. To perceive another person with the least error that virtuality will permit, a therapist must retain above all his childlike capacity for wonder, his readiness to discover something wholly astonishing under this leaf, behind this tree, or in this mind. Those who have lost this quality will find patients like Reader's Digest condensed books—where, by purging the particular, the stories are strangely identical.

The acquisition of stereotypes is not the only disadvantage of a therapist's education. Nothing kills a treatment faster than the stupefying inertness that psychotherapy training studiously cultivates. Freud's instructions: "The physician should be opaque to the patient, and, like a mirror, show nothing but what is shown to him." He commended the coldness he thought necessary in surgeons and advised his disciples that a successful therapist "pushes aside all his affects and even his humane compassion and posits a single aim for his mental forces—to carry through the operation as correctly and effectively as possible." These words formed the basis for teaching generations of prospective therapists to assume the immobility of a statue. Some of the profession's oddest moments have resulted: practitioners who balk at disclosing their marital status, refuse to shake hands with patients, and in one case, a therapist who announced to patients his policy of not laughing at their jokes.

While purists like this took him at his word, Freud's own practices ranged far from the sterility he prescribed. He had patients to dinner and developed friendships with his favorites. He treated his pal Max Eitington while strolling through the streets of Vienna. He solicited large donations to psychoanalytic causes from his wealthier clientele. He psychoanalyzed his own daughter.

Freud's enviable advantage is that he never seriously undertook to follow his own advice. Many promising young therapists have their responsiveness expunged, as they are taught to be dutifully neutral observers, avoiding emotional contact more fastidiously than a surgeon shrinks from touching an open incision with his unsterilized hand. The result is lethal. If psychotherapy were just lengthy discourse, blankness would be merely a bore. But since therapy is limbic relatedness, emotional neutrality drains life out of the process, leaving behind the empty husk of words.

SWEPT INTO THE CURRENT

A responsive therapist feels the traction of his patient's mind, and he comes to share in some of those silent emotional convictions—to *know* what the other knows. His perceptions, memories, and expectations bend in the winds of another's storm. At his best, a therapist feels this pressure *and* its wrong elements together. Then he can work to counter it, step by miniature step: *not that way*, he may say to himself or the patient, *this way*. But if the patient's mental magnet is strong or his own weak, he may be swept into the current of strange Attractors without realizing it.

Their relationship then enters the realm of traumatic repetition: patient and therapist live out whatever principles a patient's mind already contains. Then a therapist criticizes the adult who was castigated as a child, or rejects the patient a mother once abandoned, or opposes the independence of someone stifled by his father's neediness, or tramples on the accomplishments of one whose youthful talents were resented. The strength of a therapist's own Attractors, creating the power and resilience of his own emotional world, keeps him grounded, just as a mountaineer can extend a hand to a slipping climber when his own anchoring lines and pitons are strong enough to enable that daring. The therapist keeps a foot in both virtualities. If his own insides cannot resist the influence of the patient's Attractors, if his own limbic moorings are not as strong as he thinks they are, he may lose his footing and both will tumble into the patient's world.

An irony of the therapeutic process (and one unpopular with patients) is that successful therapy cannot avoid triggering the same Attractors it seeks to disarm; the patient cannot escape reliving the emotional experiences he most wishes to rid himself of. If we could hone psychotherapy to an instrument of inconceivable precision, it would still entail instances of traumatic repetition.

The only guarantee against them is an emotional distance that dooms limbic effectiveness.

WHERE WORLDS MEET

Psychotherapy is as specific as any attachment. When Lorenz imprinted goslings, they followed him but not other Austrian ethologists. A golden retriever outside a grocery store has only his owner in mind. And a patient attaches to the therapist he has.

The unsettling corollary: a therapy's results are particular to *that* relationship. A patient doesn't become generically healthier; he becomes more like the therapist. New-sprung styles of relatedness, burgeoning knowledge of relationships and how to conduct them, unthinking moves in the ballet of loving—all shift closer to those in the mind of the healer a patient has chosen.

A gathering cloud looms over the patchwork landscape of psychotherapy: the growing certainty that, despite decades of divergent rectification and elaboration, therapeutic techniques *per se* have nothing to do with results. The United States alone sports an inventive spectrum of psychotherapeutic sects and schools: Freudians, Jungians, Kleinians; narrative, interpersonal, transpersonal therapists; cognitive, behavioral, cognitive-behavioral practitioners; Kohutians, Rogerians, Kernbergians; aficionados of control mastery, hypnotherapy, neurolinguistic programming, eye movement desensitization—that list does not even complete the top twenty. The disparate doctrines of these proliferative, radiating divisions often reach mutually exclusive conclusions about therapeutic propriety: talk about this, not that; answer questions, or don't; sit facing the patient, next to the patient, behind the patient. Yet no approach has ever proven its method superior to any other. Strip away a therapist's orientation, the journals he reads, the books on his shelves, the meeting he attends—the cognitive framework his rational mind demands—and what is left to define the psychotherapy he conducts?

Himself. The *person* of the therapist is the converting catalyst, not his order or credo, not his spatial location in the room, not his exquisitely chosen words or denominational silences. So long as the rules of a therapeutic system do not hinder limbic transmission—a critical caveat—they remain inconsequential, neocortical distractions. The dispensable trappings of dogma may determine what a therapist *thinks* he is doing, what he talks about when he talks about therapy, but the agent of change is who he *is*.

That makes selecting one's therapist a life decision with (in mild terms) extensive repercussions. An uncomfortably large number of therapies yield neutral results; the only record of their existence is time spent, words spilled, and money that changed hands. But if therapy *works*, it transforms a patient's limbic brain and his emotional landscape forever. The person of the therapist will determine the shape of the new world a patient is bound for; the configuration of *his* limbic Attractors fixes those of the other. Thus the urgent necessity for a therapist to get his emotional house in order. His patients are coming to stay, and they may have to live there for the rest of their lives.

MYTHS AND MOUNTAINS

Revising limbic Attractors takes vast vistas of time—three, five years, sometimes more. People blanch when therapists speak of their profession's yawning temporal gulch. That dismay is understandable: therapy is as time-consuming and costly as a college education. But, to paraphrase Harvard's president, Derek Bok, those put off by the expense of education may find ignorance an even costlier indulgence.

Emotional perplexity exacts at least as high a life price as intellectual benightedness. Wouldn't it be fabulous if one could compress a course of limbic instruction from years into weeks or even days? The tantalizing mirage of a short (and cheap) psychotherapy, a cool and inviting oasis, has lured many across the parched sands of impossibility. The architecture of the emotional mind makes effective, fast-food therapy as much a creature of myth as the unicorn.

Psychoanalysts first explained the annoyance of therapy's requisite duration by positing *resistance:* a patient's motivated unwillingness to change, hiding like a troll under a bridge beneath his stated *desire* to change. Uncovering the iterative nature of emotional learning dispenses with that goblin, but psychotherapy's stubborn span of years remains. Limbic templates form when the brain's plasticity is fresh, when neural networks are young and malleable. By adulthood, durable Attractors roll on with the easy momentum of a bowling ball. The process capable of deflecting lives in flight operates by the progressive, painstaking transformation of one intuition into another. And so therapy consumes time. But our society has scant patience for the gradual. It keeps trying to invent instant remedies—now more than ever, when the pressure from insurance companies to sideline long-term treatments of every sort has a powerful impetus of its own.

Denying access to services, whether they are effective or not, is now the *raison d'être* of the insurance industry. Pesky legal entanglements, however, impede those carriers from a straightforward stiff-armed rebuff—patients must be discouraged, diverted, connived by gentler means. And thus insurers have taken up extolling the virtues of the shrinking morsels they are willing to provide. "Who needs psychoanalysis for eight years if you can get your needs met in 20 sessions?" trumpeted Michael Freeman, president of the Institute for Behavioral Health Care, in a 1995 *Wall Street Journal* article.

In 1995, the vaunted twenty sessions were occasionally obtainable; today, such largesse is unthinkable extravagance. Managed care offers anywhere from two to six initial sessions, but the recipient cannot know when or where it will end. Perhaps the clerk overseeing his case will grant a two- or three-session extension. Perhaps not. Every few meetings requires the filing of further reports. The treatment proceeds in convulsive fits and starts, under a perpetual pall of uncertainty incompatible with a limbic bond.

The brevity of minitherapies is another efficient forestaller of healing. The neocortex rapidly masters didactic information, but the limbic brain takes mountains of repetition. No one expects to play the flute in six lessons or to become fluent in Italian in ten. But while most can omit Ravel and Dante from their lives without sacrificing happiness, the same cannot be said of emotional and relational knowledge. Their acquisition requires an investment of time at which our culture balks.

As treatments withered under the penurious gaze of insurers, they went from minimal to functionally nonexistent. Three sessions do not differ from no sessions, except in the degree of honesty that accompanies the offer. Industry boosters backing microminitherapies may deceive the hopeful and the unwary today. If their practices continue long enough, an entire generation of patients and practitioners will forget that the treatment of people in emotional pain was ever done another way. If managed care providers were to disallow mental health treatments entirely, patients could at least be sure of what they're not getting. The current climate asks patient and therapist to wrap themselves in the emperor's new clothes and pretend that doing so will warm them both.

Despite the insignificance of complicated canons and calculated technique, all therapies are not created equal. Some are compatible with the human heart and work within its architecture to maximize health. Others, including the short and sputtering treatments now prevalent, flout limbic laws and thwart potential. That waste is painful to witness, because the limbic connectedness of a working psychotherapy requires uncommon courage. A patient asks to surrender the life he knows and to enter an emotional world he has never seen; he offers himself up to be changed in ways he can't possibly envision. As his assurance of successful transmutation he has only the gossamer of faith. At the journey's end, he will no longer be who he was, and his guide is someone he has every reason to mistrust. What Richard Selzer, M.D., once wrote of surgery is as true of therapy: only human love keeps this from being the act of two madmen.

Nine

A WALK IN THE SHADOWS

How culture blinds us to the ways of love

It is difficult
to get the news from poems
yet men die miserably every day
for lack
of what is found there.
—William Carlos Williams

The evolution of the limbic brain a hundred million years ago created animals with luminescent powers of emotionality and relatedness, their nervous systems designed to intertwine and support each other like supple strands of a vine. But in life, as on the Greek stage, every attribute confers a matching vulnerability; each heroic strength finds its mirror image in a tragic flaw. So it is with the neural skills that constitute emotional life. The limbic brain bestows experiential riches denied simpler creatures, but it also opens mammals up to torment and destruction. An alligator never feels the pain of loss, and a rattlesnake never suffers illness or death upon separation from its parents or progeny. Mammals can and do.

The neural structures responsible for emotional lives are not infinitely adaptable. Just as the dinosaur body was built to live within a range of temperature, so the limbic brain chains mammals to a certain emotional climate. The giant reptiles vanished when the skies darkened and temperatures fell. Our downfall is equally assured if we push our living conditions beyond the limits our emotional heritage decrees.

Because our minds seek one another through limbic resonance, because our physiologic rhythms answer to the call of limbic regulation, because we change one another's brains through limbic revision—what we do inside relationships matters more than any other aspect of human life. We can conduct marriages, raise children, and organize society in whatever manner we decide. Every choice (to varying degrees) suits or flouts the heart's changeless needs. An apparently straightforward and rewarding course of action can ramify into emotional predicaments that no one would deliberately select. People vary in their awareness of emotional imperatives. Those who grasp them live better lives; those who do not suffer inexorable consequences.

The same is true of larger societies. Cultures transform themselves in a few decades or centuries, while human nature cannot change at all. The likelihood of collision between cultural dictates and emotional exigencies is significant. Some cultures encourage emotional health; others do not. Some, including modern America, promote activities and attitudes directly antithetical to fulfillment.

Instead of protecting us from the frailties of the limbic brain, American culture magnifies them by obscuring the nature and need for love. The price for that failure is high. Every solid object casts a shadow, and the architecture of the emotional mind is no exception. The human heart is an early morning avenue, one half a sunny promenade where lovers walk and children play, the other side draped in velvet shade. Flowers of sadness and tragedy, and occasionally evil, grow on its darker side.

Kids These Days

An infant does his best to keep parents close: he gazes and burbles when they are near, waves and clutches if they move away, and wails into the vacuum of their absence. The infantile arsenal of enticement typically meets with unparalleled success. In a baby, parents find both miniature tyrant and artful enchanter—his tiny burps and grunts arouse bustling concern; his contentment imparts parental bliss. A baby's ability to keep parents beside him has evolved not to serve whim but limbic necessity. Eons of experience direct his brain to hold open the emotional channel that stabilizes his physiology and shapes his developing mind.

From the first hours of life, Americans traditionally sever this connection at night. Our culture assumes that a baby shouldn't sleep with his parents.

The issue of an infant's nocturnal location is reverberating around the national consciousness, thanks to the awakening of a fractious debate. Many American pediatricians frown upon cosleeping. Dr. Spock warned against the practice decades ago in his monumentally influential volume, *Dr. Spock's Baby and Child Care.* "I think it's a sensible rule not to take a child into the parents' bed for any reason," he wrote. Spock had a lighter touch than pediatrician Richard Ferber, who has waged a veritable crusade against the idea of parents and young children sharing a room or a bed.

Ferber relies on Freud's questionable habit of attributing to infants and toddlers an adult awareness of swirling sexual motives. Young children, Ferber declares, find parental sleeping accommodations "overly stimulating." He goes on to intone: "If you allow him to crawl in between you and your spouse, in a sense separating the two of you, he may feel too powerful and become worried. . . . He may begin to worry that he will cause the two of you to separate, and if you ever do he may feel responsible." The mistake here is "adultomorphism"—presuming that children are grown-ups viewed through the wrong end of a telescope: tiny, mute, fully outfitted with mature sensibilities and concerns. If a baby thought like a twenty-year-old, then perhaps he *would* suffer the ailments that, for Ferber, incontrovertibly follow a night in bed with his folks: confusion, anxiety, resentment, guilt.

On the other side of the aisle are evolutionary psychologists and cross-cultural sociologists, who point out that the American habit of sleeping separately is a global and historical singularity. Almost all the world's parents sleep with their children, and until the last sliver of human history, separate sleep was surpassingly rare. The burden of proof thus falls upon our culture to justify its anomalous nighttime practices. Robert Wright, a prominent proponent of evolutionary psychology and a champion of common sense, refutes Ferber:

> *According to Ferber, the trouble with letting a child who fears sleeping alone into your bed is that "you are not really solving the problem. There must be a reason why he is so fearful." Yes, there must. Here's one candidate. Maybe your child's brain was designed by natural selection over millions of years during which mothers slept with their babies. Maybe back then if babies found themselves completely alone at night it often meant something horrific had happened—the mother had been eaten by a beast, say. Maybe the young brain is designed to respond to this situation by screaming frantically so that any relatives within earshot will discover the child. Maybe, in short, the reason that kids left alone sound terrified is that kids left alone naturally get terrified. Just a theory.*

As Wright acknowledges, many features of the modern world, while unnatural, are not necessarily harmful. Central heating, frequent bathing, and reading glasses come to mind. Are sleeping arrangements another modern choice, neutral with respect to health? Ferber warns that the desire to share a family bed is a psychological oddity that may necessitate "professional counseling" to resolve. To support his claim, he offers recycled Freudian hash but no facts. The work of sleep researchers, however, has raised the possibility that separate sleep itself entails physiologic risk.

A baby may die suddenly and quietly asleep, and without evidence of trauma or illness, as if the soul so freshly deposited in that diminutive form was not quite fixed in place and slipped out to return to the spirit world. What parents formerly feared as crib death goes by the name sudden infant death syndrome, or SIDS. The syndrome remains mysterious. A few cases have been reclassified as covert homicides, but in the vast majority of SIDS deaths, no physical or environmental abnormality can be found. Disparate rates in different societies, however, point to a cultural contribution. Despite its advanced medical technologies and sophisticated pediatric care, the United States has the highest incidence of SIDS in the world: two deaths for every thousand live births—ten times Japan's rate, and one hundred times Hong Kong's. Yet in some countries the syndrome is virtually unknown.

Sleep scientist James McKenna and his colleagues conducted an unprecedented study that may shed light on the mystery of SIDS. They studied how babies sleep in the environment prepared for them over millions of years of hominid evolution—maternal proximity. McKenna found that a sleeping mother and infant share far more than a mattress. Their physiologic rhythms in slumber exhibit mutual concordances and synchronicities that McKenna thinks are life-sustaining for the child. "[T]he temporal unfolding of particular sleep stages and awake periods of the mother and infant become entwined," he writes. "[O]n a minute-to-minute basis, throughout the night, much sensory communication is occurring between them." Paired mothers and infants spend less time in sleep's deepest stages and have more arousals than their solitary counterparts— neural changes that, McKenna feels, protect infants from the possibility of respiratory arrest. Cosleeping infants breast-fed three times as much as solo ones and were always in the supine position, both factors that also protect against SIDS. Small wonder that the human societies with the lowest incidence of SIDS are also the ones with widespread cosleeping.

The sleep separatists exhumed the Pavlovian attitude toward

children that dominated psychology early in the last century. Reward a child's distress with attention, they said (and say today), and you increase the probability of recurrence. A child left alone at night, with no human presence to "reward" him, eventually stops crying and makes do without. But sleep is not a reflex, like the canine salivation a flank steak provokes. The dozing adult brain rises and descends through half a dozen distinct neural phases every ninety minutes, in gradually lengthening symphonic movements that culminate in morning wakefulness. Sleep is an intricate brain rhythm, and the neurally immature infant must first borrow the patterns from parents.

Infants are born knowing this—the typical baby, whether placed on his mother's left or right, spends the entire night turned toward her, with ears, nose, and occasionally eyes drinking in the sensory stimulation that sets his nocturnal cadences. The same principle allows a ticktocking clock to regularize the restless sleep of puppies newly taken from their mother, and enables the breathing teddy bear to stabilize the respirations of preemies.

Although it sounds outlandish to some American ears, exposure to parents can keep a sleeping baby alive. The steady piston of an adult heart and the regular tidal sweeps of breath coordinate the ebb and flow of young internal rhythms. Intuitive accordance with these ancient programs leads women, whether right- or left-handed, to cradle a baby in the left arm—with his head close to her heart. That laterality cannot be custom or cultural predilection, because gorillas and chimpanzee mothers show the same, innate, left-sided cradling.

The family bed debate dances around an American conundrum: we cherish individual freedoms more than any society, but we do not respect the process whereby autonomy develops. Too often, Americans think that self-rule can be foisted on someone in the way a traveler thrusts a bag at a bellhop: compel children to do it alone, and they'll learn how; do it with them and spawn a tentacled monster that knows only how to cling. In truth, premature pressure stunts the genuine, organic capacity for self-directedness that children carry within them. Independence emerges naturally not from frustrating and discouraging dependence, but from satiating dependence. Children rely heavily on parents, to be sure. And when they are done depending, they move on—to their own beds, houses, and lives.

A dog possesses no instinct to stay off the couch; if you wish him to abandon the comfort of plush pillows, you have to train him. A rat has no intrinsic desire to run a maze, but the right combination of lures and punishments can make him do so. Children need no forcing or foot shocks or food pellets to instill independence. "The one thing in the world of value," Emerson observed, "is the active soul—the soul free, sovereign, active. This every man is entitled to; this every man contains within him, although in almost all men, obstructed, and as yet unborn." That second effusion of life is the work of childhood; love and security are the patient midwives whose ministrations bring forth a new soul.

Doctors once told American women that breast-feeding was an archaic perversion. After technology produced safe and convenient bottles, mothers who *wanted* an infant at the breast were thought backward and probably pathologic. Now we know that natural feeding suits a baby's needs as no artificial substitute can. The ratio of nutrients in her milk agrees with his metabolism, and the antibodies she transfers during feeding confer lifesaving immunity against disease. Medical opposition to breast-feeding is a historical relic. This generation of mothers labors under an equally dubious pronouncement—that babies sleep best in isolation. Every infant knows better. His protest at nocturnal solitude contains the wisdom of millennia.

Surpassing the acrimonious disagreement about how to care for young children at night is the explosive conflict over the proper disposition of their daytime hours. A factual kernel is the only consensus item: young American children once spent most of their days with their mothers, and many no longer do. A spectrum of surrogates occupy modern babies and toddlers: relatives, live-in *au pairs,* regular or revolving nannies, neighbors, institutional day care workers, television shows, Disney videos, interactive computer games. Does it make a difference with whom a young child passes his time? So long as his attention is occupied and he keeps out of harm's way, does it matter whether his caretaker is a parent, a grandmother, a nanny, a stranger, an electronic device?

These questions revolve around an inconvenient center of gravity: the specificity of a child's limbic needs. If he wants only respite from boredom, any colorful distraction suffices; if he requires just the reassurance of a protective presence, any adult will do. But decades of attachment research endorse the conclusion that children form elaborate, individualized relationships with special, irreplaceable others. Investigations into the biology of that bond suggest that its preciousness emerges from neural synchrony between child and parent, and that adult neural patterns will impress themselves on a malleable brain. If so, then some of the situations in which our children currently dwell will not produce the same result as the luxuriously prolonged immersion within a small circle of devoted caretakers.

A child's electronic stewards—television, videos, computer games—are the emotional equivalent of bran; they occupy attention and mental space without nourishing. An ironic revelation of the television-computer age is that what people want from machines is humanity: stories, contact, and interaction. (Nature took a few billion years to create that kind of mechanism from scratch, so perhaps we should not wait for Silicon Valley to produce one any time soon.) Today's machines deliver not a limbic connection but imprecise simulations. Small wonder that Internet use in adults actually *causes* depression and loneliness. "We were surprised to find that what is a social technology has such anti-social consequences," said that study's author. However enticing their entertainment value, mechanical companions are unworkable relationship substitutes for adults and children alike.

Next to the role of a child's human companions. The amount and quality of love a child receives have long-lasting neural consequences. The evidential support for these contributions continues to grow brick by empirical brick: an emotional void often proves fatal to babies. Neglect produces children whose head circumferences are measurably smaller, whose brains on magnetic resonance scanning evidence shrinkage from the loss of billions of cells. Children whose mothers are depressed early in life evidence persistent cognitive deficits. Twenty years of longitudinal data have proven that responsive parenting confers apparently permanent personality strengths. Primate rearing studies have detailed the neural devastation that follows early isolation, as well as the subtler derangements that persist in a young monkey's brain from placing his mother under emotional stress. Even young rats receiving more nurturance have better developmental trajectories than their less-coddled littermates. The unimpeachable verdict: love matters in the life of a child.

All of us are born with limbic machinery running soundlessly behind wide eyes. The crucial question—just how much flexibility does that system permit? No one knows for sure. The mere presence of a single, constant caretaker cannot ensure health, since an optimal parent has to be capable and attuned. Not everyone is. With multiple caretakers, the problem divides: first, how good is *each* at tuning in to the child and regulating him? Second, when do the unavoidable discontinuities of serial caretakers prove incompatible with instilling emotional stability?

Parents and relatives wield an obvious advantage in the quality department: their exertions are spontaneous labors of love. Some nannies and day care workers feel a genuine and abiding fondness for the children they supervise. Even so, their affection does not rival parental passion. With rare exceptions, other people's children simply do not elicit the same reckless, selfless devotion that one's own offspring naturally evoke. If child care jobs commanded lavish salaries (in place of the minimum wage typically doled out), the *intrinsic* barriers to love would still be formidable. Who but an enthralled parent will attend so closely that he learns all of a child's subtle cues, picks up on the tiniest signals, and enters into the creation of a personal limbic dialect? Who else will feel the spontaneous ardor, fascination, and patience that are the requisite attendants to every complicated, creative endeavor?

Successful synchrony between two mammals is an acrobatic maneuver wherein each catches another's rhythms and adjusts his own accordingly. Parent and child are circus jugglers deftly exchanging bowling pins; a stranger cannot smoothly step into their practiced rhythm without disruption. The hard truth is that few employees will feel inspired to *love* someone else's child in the fullest sense of the word—and even if they did, the task would be exceedingly difficult. To make matters worse, the average out-of-home care provider faces not the complex emotional requirements of *one* customer's child but, at best, three or six at a time. Economies of scale exist for clear reasons, but the impersonal element they inject is anathema to a baby's developing emotional needs. To paraphrase Mark Twain: the difference between a caretaker who tunes in to a child and one who almost tunes in is as great as the difference between lightning and a lightning bug.

All other things being equal—and sometimes they are not— parents stand the best chance of meeting young limbic needs because of the regularity of their presence and the natural depth of their devotion. Inverting those advantages exposes the shortcomings of institutional day care. An infant's brain is designed for ongoing attunement with the people predisposed to find him the most engaging of all subjects, the most breathtakingly potent axis around which their hearts revolve. Instead he finds himself competing with half a dozen peers for the emotional focus of an understandably unimpassioned surrogate. Will a parent's infatuation and the divided attentions of sequential strangers exert a comparable effect on the developing brain? The pull to believe so emanates from wishful thinking rather than the plausibility of the hypothesis itself.

How much affectionate parental momentum does a baby need, then, to keep him coasting through impersonal limbic flatlands? Love is a bodily process and therefore consumes an incompressible amount of time. Each week a child's extrafamilial hours can range only from 0 to 168, the numerical limits to *never* and *always.* Between those temporal bookends, a certain number must exist—call it X hours—where the impact of parental absence shades from negligible into emotional risk.

Most agree that X, varying somewhat from child to child, increases with age—babies need togetherness most, toddlers a bit less, older children still less. For any particular age, where does it fall? Does X equal five, ten, twenty, forty, eighty hours per week? The equation does not lend itself to easy study, owing to the exuberant proliferation of confounding variables and the equally confusing exaggerations and distortions perpetuated by a number of parties to the debate. Several researchers have shown that day care in excess of twenty hours per week for children under a year old increases the risk of insecure attachment and its negative emotional effects. Dissenting studies found that extensive day care inflicts no discernible influence when children receive the high-quality variety: big budgets, competent staff, favorable adult-to-child ratios, low turnover—conditions remote from the actual terms under which out-of-home care operates in this country.

With its high percentage of single-parent families and double-income households, discovering the value of X is a pursuit our culture cannot afford to delay. Unfortunately, the slow road to postponement is precisely the path we are on. Such vitriolic controversy erupted over the initial day care data that it choked off meaningful scientific debate. Psychologist Robert Karen described the difficulty, while gathering material for a book, in getting researchers to voice *any* audible viewpoint. When day care came up, otherwise voluble scientists went mum, apparently intimidated by political heat. "I wouldn't have any opinion," one developmental researcher told Karen (if *he* doesn't, who does?), and when the latter pressed him, amended his position from incredible to sadly believable: "I wouldn't have one for publication."

All cultures, postdating the advent of science, have protected certain propositions from empiricism's cold eye because even the possibility of an unfavorable result is deemed unacceptable. In this time and place, our culture promotes as self-evident the notion that employable adults must have jobs and careers, that children do fine with less. The extent of cultural reliance on this cornerstone is unrelated to its ultimate soundness, which the science of some future age will assess. "You can fight for a cause and make it come about," writes William Gass, "but you can never make an idea come true like a wish, for its truth is—thank heaven—out of all hands." In modern America, ignorance of the developmental extent of parental love is perilous. *Choosing* ignorance is begging for trouble. If we ask a parent to consider that modern lifestyles may deprive his child of a vital limbic ingredient, a neural vitamin, an emotional vaccine against later illness—then we risk arousing guilt and distress. If we leave the question alone as untouchable, and parents unknowingly shortchange their children, everyone will feel worse.

These matters are irrelevant to parents without any choice—just as information on optimal diets is useless to those who cannot afford food. But some parents do have some temporal leeway, and others may have more than they think. Many parents, particularly mothers, find it excruciating to leave young children behind for days at a time. Limbic pain of that magnitude should not be dismissed as a trifle without the most careful deliberation, the best possible evidence. Parents who contemplate staying at home to raise children are treated instead to the cultural chorus of well-meaning dissuaders: you're bright; you're talented; wouldn't you rather *do* something with your time? The implication is clear: love doesn't accomplish; it does nothing we need done. In its baldest tally of values, our culture automatically equates a dedication to full-time parenting with the absence of ambition. But in what human activity could there possibly be *more*?

Government is the machine that transforms cultural attitudes into policies. "What a society honors," wrote Aristotle, "will be cultivated." We need only glance at our social programs to catch the direction of prevailing winds. On one side, conservatives dismantle welfare so that single mothers must set children aside and return to work—not the labor of raising children, but the *real* work our culture values and upholds. On the other hand, liberals champion child care initiatives calling for an expansion in institutionalized surrogate care. Caught in the middle, American parenthood is beleaguered, belittled, and besieged.

Because so much of parenthood is giving—time, attention, patience, food, guidance, love—*incoming* emotional sustenance is indispensable for balance. The presence of two parents is neither unintended superfluity nor mere economic advantage; parents *need* each other for support and replenishment. And yet in more and more families, one adult bears the lopsided burdens alone. A third of American children grow up in mother-only households; one half will live in a single-parent home before turning eighteen. The outcome of such arrangements is predictable: parents who receive inadequate love have less to give—to anybody, including their children. Psychologist Judith Wallerstein studied families for five years after divorce and found in the children "a dismayingly high incidence of depression" at every point along the way. Divorce's great danger for children is, she writes, "in the diminished or disrupted parenting which so often follows in the wake of the rupture and which can become consolidated within the postdivorce family."

Humanity's limbic ties make any social structure a web of interdependencies, wherein perturbations ripple and reverberate, inward and out. Children cannot bloom in a culture where adults do not cohere long enough so that intact families may support needful hearts. And youngsters who grow up without knowing the fullness of love will be fighting the odds when they mount their own struggle to establish a life bond with another. The emotional fate of children is inextricably bound to the ability of their parents to love one another—a skill that is falling into disrepair.

THE DYNAMICS OF DUOS

Relationships, including romantic ones, are fantastic concatenations of limbic energies. Humanity has had eons to become familiar with these ancient forces, but today it seems we apprehend their essence less than ever. Loving relationships plainly perplex our limbically incognizant culture. Bookstores bulge with so many how-to relationship primers that it seems nobody knows how. That ignorance extracts a painful pound of flesh. "Fathers and teachers," wrote Dostoyevsky, "I ponder the question, 'What is Hell?' I maintain it is the suffering of being unable to love." Too many of our citizens spend their lives in that purgatory, searching vainly for a redemption that eludes them. What don't they know? What doesn't our culture teach them?

The simple equations of love. Like this: *relationships live on time.* They devour it in the way that bees feed on pollen or aerobic cells on oxygen: with an unbending singularity of purpose and no possibility of compromise or substitution. Relatedness is a physiologic process that, like digestion or bone growth, admits no plausible acceleration. And so the skill of becoming and remaining attuned to another's emotional rhythms requires a solid investment of years.

Americans have grown used to the efficiencies of modern life: microwave ovens, laser price scanners, number-crunching computers, high-speed Internet access. Why should relationships be any different? Shouldn't we be able to compress them into less time than they took in the old days, ten or a hundred or a thousand years ago? The unequivocal limbic *no* takes our culture by surprise. The modern American is genuinely puzzled when affiliations evaporate from inattention. Every new second of togetherness reestablishes the terms of a relationship. But cultural mythology imbues social ties with the clumsy durability of *things*—once attained, always attainable; once established, easy to get back to weeks, months, years later. The truth is only slightly less dire than the words of playwright Jean Giraudoux: "If two people who love each other let a single instant wedge itself between them, it grows—it becomes a month, a year, a century; it becomes too late."

Some couples cannot love because the two simply don't spend enough time in each other's presence to allow it. Advances in communication technology foster a false fantasy of togetherness by transmitting the impression of contact—phone calls, faxes, e-mail—without its substance. And when a relationship is ailing from frank time deprivation, both parties often aver that nothing can be done. Every activity they spend time on (*besides* each other) has been classified as indispensable: cleaning the house, catching the news, balancing the checkbook.

Such an existence is too expensive to bear. When launching a life raft, the prudent survivalist will not toss food overboard while retaining the deck furniture. If somebody must jettison a part of life, time with a mate should be last on the list: he needs that connection to live.

Couples do not receive this advice from friends, colleagues, family—their world. Instead they are encouraged to achieve, not attach. Americans spur one another to accomplish and acquire before anything else—our national dream holds that industry leads to a promised land, and nobody wants to miss out on a share of paradise. When consummating a career does not bring happiness—as it cannot—few pause to reconsider their assumptions; most redouble their efforts. The faster they spin the occupational centrifuge, the more its high-velocity whine drowns out the wiser whisper of their own hearts.

When they do get down to relating, Americans find they have been tutored for years in the wrong art. In a dazzling vote of confidence for form over substance, our culture fawns over the fleetingness of being *in love* while discounting the importance of *loving.*

A child tunes in to the emotional patterns of parents and stores them. In later life, if he spots a close match, the key slides in the psychobiologic lock, the tumblers fall home, and he falls *in love.* The accuracy of limbic architecture astounds. In a city of 5 million people, in a country of 270 million, in a world of 6 billion, people pick partners emotionally identical to their predecessors and swoon. *In love* twists together three high-tensile strands: a potent feeling that the other fits in a way that no one has before or will again, an irresistible desire for skin-to-skin proximity, and a delirious urge to disregard all else. In the service of that prismatic blindfold, *in love* rewrites reality as no other mental event can. "Whoso loves," wrote Elizabeth Barrett Browning, "believes the impossible."

Our society goes the craziness of *in love* one better by insisting on the supremacy of delectable but ephemeral madness. Cultural messages inform the populace that if they aren't perpetually electric they are missing out on the pinnacle of relatedness. Every pop-cultural medium portrays the height of adult intimacy as the moment when two attractive people who don't know a thing about each other tumble into bed and have passionate sex. All the waking moments of our love lives should tend, we are told, toward that throbbing, amorous apotheosis. But *in love* merely brings the players together, and the end of that prelude is as inevitable as it is desirable. True relatedness has a chance to blossom only with the waning of its intoxicating predecessor.

Loving is limbically distinct from *in love.* Loving is mutuality; loving is synchronous attunement and modulation. As such, adult love depends critically upon *knowing* the other. *In love* demands only the brief acquaintance necessary to establish an emotional genre but does not demand that the book of the beloved's soul be perused from preface to epilogue. *Loving* derives from intimacy, the prolonged and detailed surveillance of a foreign soul.

People differ in their proficiency at tracing the outlines of another self, and thus their ability to love also varies. A child's early experience teaches this skill in direct proportion to his parents' ability to know *him.* A steady limbic connection with a resonant parent lays down emotional expertise. A child can then look inside someone else, map an emotional vista, and respond to what he senses. Skewed Attractors impair a person's ability to love freely and well. His heart's gaze, in the manner of one whose eyes do not properly focus, will have the unsettling habit of looking beyond and behind the person in front of him. A heart thus displaced falters in its efforts to meet another's rhythms, to catch another's tempo and melody in the duet of love.

Because *loving* is reciprocal physiologic influence, it entails a deeper and more literal connection than most realize. Limbic regulation affords lovers the ability to modulate each other's emotions, neurophysiology, hormonal status, immune function, sleep rhythms, and stability. If one leaves on a trip, the other may suffer insomnia, a delayed menstrual cycle, a cold that would have been fought off in the fortified state of togetherness.

The neurally ingrained Attractors of one lover warp the emotional virtuality of the other, shifting emotional perceptions— what he feels, sees, knows. When somebody loses his partner and says a part of him is gone, he is more right than he thinks. A portion of his neural activity depends on the presence of that other living brain. Without it, the electric interplay that makes up *him* has changed. Lovers hold keys to each other's identities, and they write neurostructural alterations into each other's networks. Their limbic tie allows each to influence who the other is and becomes.

Mutuality has tumbled into undeserved obscurity by the primacy our society places on the art of the deal. The prevailing myth reaching most contemporary ears is this: *relationships are 50-50.* When one person does a nice thing for the other, he is entitled to an equally pleasing benefit—the sooner the better, under the terms of this erroneous dictum. The physiology of love is no barter. Love is simultaneous mutual regulation, wherein each person meets the needs of the other, because neither can provide for his own. Such a relationship is *not* 50-50—it's 100-100. Each takes perpetual care of the other, and, within concurrent reciprocity, both thrive. For those who attain it, the benefits of deep attachment are powerful—regulated people feel whole, centered, alive. With their physiology stabilized from the proper source, they are resilient to the stresses of daily life, or even to those of extraordinary circumstance.

Because relationships are mutual, partners share a single fate: no action benefits one and harms the other. The hard bargainer, who thinks he can win by convincing his partner to meet his needs while circumventing hers, is doomed. Withholding reciprocation cripples a healthy partner's ability to nourish him; it poisons the well from which she draws the sustenance she means to give. A couple shares in *one* process, *one* dance, *one* story. Whatever improves that *one* benefits both; whatever detracts hurts and weakens both lives.

Modern amorists are sometimes taken aback at the prospect of investing in a relationship with no guarantee of reward. It is precisely that absence, however, that separates gift from shrewdness. Love cannot be extracted, commanded, demanded, or wheedled. It can only be given.

A culture wise in love's ways would understand a relationship's demand for time. It would teach the difference between *in love* and *loving;* it would impart to its members the value of the mutuality on which their lives depend. A culture versed in the workings of emotional life would encourage and promote the activities that sustain health—togetherness with one's partner and children; homes, families, and communities of connectedness. Such a society would guide its inhabitants to the joy that can be found at the heart of attachment—what Bertrand Russell called "in a mystic miniature, the prefiguring vision of the heaven that saints and poets have imagined."

The contrast between that culture and our own could not be more evident. Limbic pursuits sink slowly and steadily lower on America's list of collective priorities. Top-ranking items remain the pursuit of wealth, physical beauty, youthful appearance, and the shifting, elusive markers of status. There are brief spasms of pleasure to be had at the end of those pursuits—the razor-thin delight of the latest purchase, the momentary glee of flaunting this promotion or that unnecessary trinket—pleasure here, but no contentment. Happiness is within range only for adroit people who give the slip to America's values. These rebels will necessarily forgo exalted titles, glamorous friends, exotic vacations, washboard abs, designer everything—all the proud indicators of upward mobility—and in exchange, they may just get a chance at a decent life.

TRUTH AND CONSEQUENCES

Just before the stroke of midnight, while Ebenezer Scrooge is having his last discussion with the Spirit of Christmas Present, he spies two skeletal figures huddling under the latter's green robes. They are the two children of Man, the Spirit tells him, Ignorance and Want. Appalled at their pitiful and cadaverous condition, Scrooge demands to know what can be done for them. "Have they no refuge or resource?" he cries. The Spirit counters: "Are there no prisons?" he asks, skewering Scrooge with his own miserly incantation against beggars. "Are there no workhouses?" The clock strikes twelve, and Scrooge faces his last and least forgiving ghost, the specter of things yet to come.

Like those who made Scrooge recoil, the dark children of our age are difficult, even painful to look upon. When a society foils limbic mechanisms, it unleashes a string of afflictions whose pathognomonic patterns are part of daily American life.

When the Centre Cannot Hold

Without rich limbic resonance, a child doesn't discover how to sense with his limbic brain, how to tune in to the emotional channel and apprehend himself and others. Without sufficient opportunity for limbic regulation, he cannot internalize emotional balance. Children thus handicapped grow up to become fragile adults who remain uncertain of their own identities, cannot modulate their emotions, and fall prey to internal chaos when stress threatens.

Anxiety and depression are the first consequences of limbic omissions. The dominant emotion in early separation's protest phase is nerve-jangling alarm. If isolation stretches out, a mammal descends into lethargic despair, depression's alter ego. Emotional disconnection produces young rhesus monkeys who suffer a lifelong vulnerability to the twin aboriginal states of nervousness and depression; the same is true in our own primate society. Close early relationships instill a permanent resilience to the degenerative influence of stress, while neglect sensitizes children to those effects. The brains of insecurely attached children react to provocative events with an exaggerated outpouring of stress hormones and neurotransmitters. The reactivity persists into adulthood. A minor stressor sweeps such a person toward pathologic anxiety, and a larger or longer one plunges him into depression's black hole.

These two emotional states are epidemic. Depression and anxiety each cost America more than $50 billion per year, colossal sums that only hint at the mountain of suffering behind them. And the mountain is growing: the rate of depression in the United States has been rising steadily since 1960. Suicide rates for young people have more than tripled over that time; killing oneself is now a leading cause of death in adolescence. Reports on child welfare detail their nutritional status, their lead exposure, the design of the straps in their car seats, but omit mention of the stability and quality of love in their lives. We would be wise to pay as much attention to those relationships as we do to the vegetable content of school lunches.

A person who lacks a stable center feels an urgent need to fill the gap—he needs *something* to orient himself as he tries to navigate the world. Since he cannot use the limbic tools that penetrate to the core of self and others, he will look to external cues—those he *can* be sure about. Thwarted attachment and limbic disconnection thus encourage superficiality and narcissism. People who cannot *see* content must settle for appearances. They will cling to image with the desperation appropriate to those who lack an alternative. In a culture gone shallow, plastic surgery supplants health; photogenicity trumps leadership; glibness overpowers integrity; sound bites replace discourse; and changing what *is* fades before the busy label-swapping of political correctness. When a society loses touch with limbic bedrock, spin wins. Substantive aspirations inevitably suffer.

If the attachment fabric of a civilization frays, if people cannot get from their relationships the emotional regulation that those bonds were designed to furnish, they will commandeer whatever means of limbic modulation they can lay hands on. Their hungering brains will seek satisfaction from a variety of ineffectual substitutes—alcohol, heroin, cocaine, and their cousins. As a society produces more people who lack access to the neural process that engenders emotional balance, the ranks of street drug users will grow.

Periodic media dispatches keep us posted about the status of America's war on drugs. But the real battle our country fights is not against drugs *per se* but limbic pain—isolation, sorrow, bitterness, anxiety, loneliness, and despair. Our culture is losing this struggle so badly that millions try to manipulate the delicate organs of emotion into delivering a respite from their daily dysphoria. If these pharmacologic experiments met with success—if they actually increased happiness and emotional fulfillment—who could object? But homespun chemical remedies for emotional pain are disastrous. Mood-altering agents sold on the street obliterate anguish for a few minutes or hours, and then they dissipate, leaving behind a deeper ache. Repeated use ravages the nervous system and undermines already desperate lives.

America's antidrug czars have encouraged us to believe that addiction exists because street drugs fasten on to the average mind like the giant *kraken* of seafaring legend: one tangential brush with this vile sucking beast and a healthy kid's mind disappears into the black depths forever. The evidence refutes this notion. Contemporary chemists (some in basements and barns) have concocted potent pharmacological snares and have enhanced the power of native ones. But examine the figures for cocaine, thought to be the most powerfully addictive substance known. Of all humans who try co-caine, less than 1 percent become regular users—the other 99 walk away. As Malcolm Gladwell has argued, this staggering imbalance points to a problem not in the juices of coca leaves, but inside the brains of the tiny fraction who find its effect on their emotions ir-resistible. America expends billions to protect our borders against the influx of small packets of limbic anesthesia. Those sums might be better spent ensuring that our children harbor brains minimally responsive to such agents.

Inherited temperament can facilitate both the willingness to try drugs and the ease of becoming dependent upon them. Neuro-science may someday offer a remedy; other than funding basic re-search, the average citizen can do little about that end of the problem. But research also indicts nurture, nature's coconspirator in all neural matters, in modulating children's vulnerability to drugs. Study after study has shown that children with close famil-ial ties are far less likely to become entangled in substance abuse. Even under ideal circumstances, teenage years abound in emotional surges, changing roles, growing pains. If adolescents do not receive limbic stability from relationships in the home, they will be mea-surably more susceptible to chemical options outside.

Debates on solving America's drug epidemic typically alternate between conservatives demanding longer prison sentences and lib-erals calling for more treatment programs. Both sides are reluctant to admit that neither approach has come anywhere near to ridding this country of our gargantuan problem. Consigning users to a penal system that combines a plentiful supply of drugs and the in-centive to use them is not a convincing prescription for ameliora-tion. Treating addiction has proved substantially more effective, when legislators are in a mood to fund it—and, since addicts lack lobbyists, that is not very often.

Cultural awareness of limbic principles informs the kind of

measures we expect to prevail in the war on drugs. Will lectures on the evils of chemical dependency deter teenagers from a life of substance dependence? Don't believe it. While their end is worthy, such talk targets the neocortical brain, not the limbic one. Pain is too potent a motivator for facts to undo. Pretending otherwise is a threadbare illusion convincing only to those who already feel basically well. The insouciance of Just Say No assumes that the human brain and will are separable. They are not. Limbic instability undermines the neural capacity for resolve that jaunty slogans call upon.

An addict's bulwark against relapse involves more communion than cogitation, as Alcoholics Anonymous and its multiple variants demonstrate. Gathering like people together to share their stories imbues a wordless strength, what Robert Frost called in another context "a clarification of life—not necessarily a great clarification, such as sects and cults are founded on, but a momentary stay against confusion." The limbic regulation in a group can restore balance to its members, allowing them to feel centered and whole. But even the solidity of the omnipresent clan is no panacea. Too often users return to the ephemeral reprieve that drugs offer.

Incarceration and treatment have their shortcomings. Prevention is the ringing alternative—not television ads and exhortatory pamphlets, but the precious prophylaxis that nature provides, in the form of a family's early love. Raising children attentively, thoroughly, and patiently immunizes their brains against stress like Salk's potion protects their bodies from polio. Love is and will always be the best insurance against the despair for which street drugs are the obvious antidote.

Limited Partnership

The limbic brain makes mammals as ready to fasten to their fellows as LEGO blocks. People form lasting attachments to a variety of others: husbands, wives, children, friends, alma maters, the baseball team nearest their home, the corporation they work for. Who hasn't felt the twinge of loyalty to an old car when it comes time to sell, or to a pair of jeans past their prime? Like Lorenz's goslings, people sometimes bond to objects incapable of reciprocating. Detached from human relationships, limbic proclivities can hobble. If a person's brain targets an emotionally inert would-be partner, attachment needs can propel him into contact with what *cannot* satisfy him, like a moth battering its wings against a streetlamp in the soft summer night. Mammals can see a deceptive light inside the inanimate, a false attachment wherein the inferred give-and-take never materializes.

Today's most treacherous false attachment springs up between human beings and corporations. In this era of downsizing and its euphemistic equivalents, the tale of the dedicated worker abruptly terminated after years of loyal service has become archetypal. Behind the stark outlines of the tale are thousands of people who pour their hearts into jobs, give beyond their monetary recompense out of team spirit, and later are unceremoniously dumped. Many such people are waylaid by the attachment mechanisms that should promote well-being but trap them instead.

Natural limbic inclinations include loyalty, concern, and affection. "When you love," wrote Ernest Hemingway, "you wish to do things for. You wish to sacrifice for. You wish to serve." Within their designed environment—a family—these impulses make fertile ground wherein healthy relatedness takes root and grows. The workplace bears strong resemblance to the home—indeed, for most of humanity's history, the work environment *was* the home. In both settings, one encounters amiable companions, authoritative overseers, shared travails.

A great divide, however, separates corporation and kin. Attachment urges prompt exploitation because companies do not have emotional impulses, and human beings do. A company has no limbic structure predisposing it to recognize *its own* as intrinsically valuable. People who extend fidelity and fealty to a corporate entity—legally a person and biologically a phantom—have been duped into a perilously unilateral contract.

Steeped as they are in limbic physiology, healthy people have trouble forcing their minds into the unfamiliar outline of this reptilian truth: *no intrinsic restraint on harming people exists outside the limbic domain.* Preparing soldiers for combat involves not only teaching them physical skills necessary to vanquish opponents but also indoctrinating the emotional outlook that creates an Enemy. That psychological goal is achieved by severing mental bonds between Us and Them while simultaneously strengthening intragroup ties. *The Enemy is not like us,* both sides tell prospective combatants, *they are subnormal, inhuman, less than animals.* The average infantryman fights not for lofty political ideals, but because homicidal fiends threaten him and the family of buddies with whom he has labored, suffered, and loved. History brims with the brutality that flows between groups when no limbic tie unites them.

Corporate malfeasance shocks many, but corporations operate outside attachment as surely as armies do. Misdeeds—even savagery—are inevitable. When the tobacco industry delivers death more efficiently than any war machine in history, it does so to *our own* people—because *our own* is a limbic, not a corporate, precept. When the Johns Manville Corporation covered up the lethal effects of asbestos, the company sent to their unknowing deaths not strangers, but hundreds of their employees. Any reptile would have done the same. Assuming mutuality where none exists is a mammal's grave and occasionally fatal error. In the Manville litigation, Charles H. Roemer, former chairman of the Paterson Industrial Commission, recounted a luncheon meeting he had with Manville's president, Lewis Brown, and his brother Vandiver Brown,

Manville's corporate attorney. The latter ridiculed other asbestos manufacturers for their foolishness in notifying workers about the terminal illness they had contracted on company time. Mr. Roemer's testimony: "I said, 'Mr. Brown, do you mean to tell me you would let them work until they dropped dead?' He said, 'Yes. We save a lot of money that way.' "

The urge to embed oneself in a family—to hold an endeavor in common with others, to be part of a team, a band, a group that struggles together toward a common victory—is an indomitable aspect of the human mind and brain. In a culture whose members are ravenous for love and ignorant of its workings, too many will invest their love in a barren corporate lot, and will reap a harvest of dust.

ALL WE NEED OF HELL

Because limbic regulation between parent and child directs neurodevelopment, social contact is necessary for evolving pieces of behavior to assemble into a functioning animal. Without parental guidance, neurochemical disjunctions accumulate and budding behaviors conglomerate into a mess. Isolation-rearing in rhesus monkeys produces a nightmarish head-banging, eye-gouging mutant with scant resemblance to a healthy, coherent organism. Rhesus monkeys must be mothered even to eat or drink in the normal simian style.

Primate experiments in isolation-rearing have lessons to teach us. Aggression is a formidably complex behavior requiring precise neural control—too little hostility impairs individual survival and too much prohibits the successful cohabitation that social animals crave. In rhesus monkeys nurtured by parents, neuroscientists have established correspondences between aggressivity and brain levels of coordinating neurotransmitters. A normal brain hums with thousands of these subtle rhythms, as microscopic machinery composes and harmonizes behaviors. Monkeys deprived of early limbic regulation have lost both neural organization and the capacity for modulated aggression. They are erratically, unpredictably, chaotically vicious. The condition is irremediable, even with the benefit of today's advanced neural pharmacopoeia. Gary Kraemer makes the chilling observation that isolation-reared monkeys do not "conform to the usual neurobiological rules any more than they conform to the usual social rules. . . . [I]t is unlikely that tinkering with increasingly specific pharmacological fixes of what seems to be a general disorganization of brain function will be successful."

Because mammals need relatedness for their neurophysiology to coalesce correctly, most of what makes a socially functional human comes from connection—the shaping physiologic force of love. Children who get minimal care can grow up to menace a negligent society. Because the primate brain's intricate, interlocking neural barriers to violence do not self-assemble, a limbically damaged human is deadly. If the neglect is sufficiently profound, the result is a functionally reptilian organism armed with the cunning of the neocortical brain. Such an animal experiences no compunctions about harming others of its kind. It possesses no internal motivation not to kill casually from minor frustration or for minimal gain. One young offender who crippled his victim during a mugging accounted for his actions in this way: "What do I care? I'm not her."

America produces remorseless killers in bulk. One hundred years ago, Jack the Ripper riveted the attention of the Western world by doing away with five people. This culture would barely notice such modest exploits—so many have surpassed the quaintly amateurish Ripper that we cannot remember their names, much less their crimes. Squadrons of soulless assassins do not germinate by chance. These avenging Phoenixes arise from the neural wreckage of what once could have been a healthy human being.

As conditions worsen, the violence emerges younger and younger. Our culture now teems with lethal children. In Colorado, a pair of teenage boys armed with bombs and automatic weapons methodically execute a dozen classmates, a teacher, and then themselves. In Arkansas, a thirteen-year-old and an eleven-year-old boy coolly ambush and gun down a group of children and teachers—five dead, ten wounded. A California court convicts a twelve-year-old of beating a toddler to death. In Oakland, a six-year-old child breaks into an apartment and kicks in a baby's skull. He is indicted for attempted murder—the youngest person ever to be charged with that crime in our country's history. All too horribly real, these events evoke national hand-wringing, confusion, despair.

Stories like these contain tragedy to spare, but less mystery than many suppose. Limbic deficits engender uncontrolled viciousness, through a process established long ago within our fragile physiology. Recall that the brains of neglected children show neurons missing by the billions. Lest anyone think those vanished cells are inconsequential, our own children prove otherwise.

As Winston Churchill observed, there is no finer investment for any community than putting milk into babies. The potential for humanity lives inside every infant, but healthy development is an effort, not a given. If we do not shelter that spark, guide and nurture it, then we not only lose the life within but we unleash later destruction on ourselves.

PRIMARY SCARE PROVIDERS

The trends in many of America's institutions are disquieting, if not alarming. Education is under siege, as classrooms in the world's wealthiest nation regularly yield a bumper crop of scholastic ineptitude. Crowded courts have slowed almost to stasis, halting the revolution of their weary wheels every now and again to issue judgments increasingly distant from any commonsense conception of justice. Politicians are so mired in raising money for the chance to denounce one another on television that government has declined into a frantic auction of favors. And so it goes. Tendrils of causation from the limbic domain touch upon all of these tendencies.

One of the most ailing American institutions is, ironically, the ancient mission of healing the sick. The last century saw a two-part transformation in the practice of medicine. First, an illness beset the relationship between doctor and patient, then radical restructuring attacked the residual integrity of that attenuated tie.

The initial stage of medicine's remodeling was the inadvertent distancing of doctors from human affairs. The first half of the twentieth century brought dazzling technologies—antibiotics, vaccines, X rays, anesthesia—that delivered sophisticated diagnostic acumen and unparalleled ability to cure. The age ushered in was also one of estrangement from patients. For the last thirty years, the paradox of Western medicine has been the seemingly inexplicable coexistence of technical excellence with unpopularity. Americans receive the world's most advanced treatments, biomechanical miracles in power and scope. Yet patients complain fiercely. Doctors don't listen, patients say; they are cold and busy technocrats. And the patients are right, because American medicine has come to rely on intellect as the agency of cure. The neocortical brain has enjoyed a meteoric ascendancy within medicine even as the limbic star has fallen into disfavor.

What doctors once knew, but cast aside for an extended gadgetry fling, is that patients come looking for both healer and expert. Illness arouses the ancient attachment machinery; it awakens a limbic need. When they go to the doctor, patients hope for the illuminating test, the correct diagnosis, the appropriate remedy. They also want someone who connects with them in spite of their suffering; they wish for a warm hand on their shoulder and the security of speaking with one who has been through this before. A dying patient described it this way:

I wouldn't demand a lot of my doctor's time. I just wish he would brood on my situation for perhaps five minutes, that he would give me his whole mind just once, be bonded with me for a brief space, survey my soul as well as my flesh to get at my illness. . . . . I'd like my doctor to scan me, to grope for my spirit as well as my prostate. Without such recognition, I am nothing but my illness.

Western medicine dismissed these tools of healing as expendable hand-holding, a luxury that busy schedules could not permit. "Bedside manner" became a cursory interchange thought mildly reassuring but inessential, particularly when compared to the *real* science of pathophysiology.

Medicine lost sight of this truth: attachment *is* physiology. Good physicians have always known that the relationship heals. Indeed, good doctors existed before any modern therapeutic instruments did, in the centuries when the only prescriptions were philters deriving their potency from metaphoric allusion to the healer's own person. The extraordinary results of the lab tests and procedures, the mastery they provided over the wily enemy of disease, proved seductive. Western medicine embraced the effective machines and ceded its historic soul.

The wholesale desertion of limbic attentiveness, once as much a part of medicine as the stethoscope, has been costly. A 1994 proposal in *The Lancet,* Europe's most respected medical journal, advocated teaching acting techniques to medical students. The proposed utility of adding theatrical training to the curriculum? Providing physicians with the means to feign concern for patients, since their incapacity to care is too embarrassingly evident. "We would suggest that if physicians do feel antipathy [toward patients], they should at least act as if they cared," wrote the medical thespians.

Here our finest doctors endeavor, without irony or shame, to pass off a good relationship as a kind of performance art. Their immodest proposal aptly captures the emptiness at the core of Western medicine. For many years, the medical community hasn't believed that anything substantive travels between doctor and patient unless it goes down a tube or through a syringe. The rest can be comfortably omitted or conveniently faked.

Patients (mammals that they are) sensed the limbic void in American medicine and deserted *en masse*. Even while traditional medicine eschewed emotional aspects of healing, multiple groups sprang up to accommodate them: acupuncturists, chiropractors, masseuses, body workers, reflexologists, herbal therapists, and a host of others. The "alternative" healers proliferated in response to the demand for a context of *relatedness*. These limbically wiser settings are friendlier to emotional needs—they involve regular contact with someone who participates in close listening, and often, the ancient reassurance of laying on hands. Alternative medicine sees these activities as quintessential rather than incidental to healing. The result? Patients vote with their feet and their wallets. Alternative medicine now collects more out-of-pocket dollars than does its traditional, shortsighted predecessor.

So deep is the divide between neocortical and limbic medicine that it extends to the pills people are willing to take. The warm welcome given alternative practitioners has emboldened manufacturers to market alternative *medications*—so-called neutraceuticals, herbal or natural remedies for ailments ranging from AIDS to menopause. What is the appeal of natural pills? People feel able to trust them, even if the efficacy of the chemical constituents is more mythic than real. Current laws do not require proof of effectiveness for food additives, the legal class to which herbal remedies belong. Indeed, regulatory laxity allows such products to omit the ingredients they list on the label. (One study of ginseng preparations, for instance, found only half included ginseng at all; only a quarter provided any in a biologically usable form.) The bustling neutraceutical business—now with $5 billion per year in U.S. sales—is an economic testament to the depth of yearning for an earlier, more trustworthy, more humane brand of medicine.

Medicine's movement away from limbic considerations abruptly accelerated in the 1990s, as solo practitioners and fee-for-service physicians congealed into the large corporate mass known as managed care. The emotional revamping was drastic: medicine was once mammalian and is now reptilian.

The administrative framework of medicine formerly permitted at least the possibility of human relationships between the participants, even if technology tended to get in the way. But the corporate takeover of the doctor-patient relationship fatally compromised medicine's ailing emotional core. A corporation has customers, not patients; it has fiscal relationships, not limbic ones. Like crocodiles incapable of an aversion to cannibalism, HMOs prosper whether or not customers are consumed in the process. Individual doctors can care about patients, but all too often they do not have authority to implement the decisions that could protect those patients from harm. In today's market, *ER* meets *Jurassic Park*. *"Caveat emptor"* has given way to *"Horrescat emptor"*—let the buyer be scared.

Corporate medicine cloaks its reptilian scales in warmer garments. Every fall, as open enrollment draws near, one cannot escape inundation in the voluptuous deluge of claimed corporate nurturance. Television and radio ads paint various insurers as hybrids of Marcus Welby, June Cleaver, and Mother Teresa. But the mother pitched never materializes. As patients have learned the hard way, HMOs and managed care outfits profit by spending less than subscribers pay in. They pursue this end with efficiency and ruthlessness. Doctors are bribed and bullied into not treating patients, while service rationers sequester themselves behind a thicket of bureaucracy so dense that it thwarts all but the most tenacious self-advocates. Many physicians are reluctant to air their discussions, but in private they are bursting with tales of corporate abuse. If you think that patients are not falling ill from preventable diseases, losing organs and limbs to deliberate delay, and dying from systematic inattention, then think again.

A Kentucky physician at a managed care organization confessed to causing a patient's death by denying him the operation he needed to live. She had feared for her job if she approved the procedure; after she made the correct corporate decision, she was promoted for administrative thrift. "The distance made it easier," she said, "like bomber pilots in war who never see the faces of their victims." The New York State Health Commissioner discovered a short while ago that an HMO was using its data on cardiac surgery death rates to improve the selective routing of patients to New York hospitals. Did the insurer send its patients to the most dependable institutions? Of course not. Instead, administrators used the statistics to bargain for basement-rate prices from the most lethal centers. Then, coaxing along cherished dividends, they diverted patients to the cheapest facilities available, where those customers were likeliest to suffer and die.

Before it is safe to go back to the doctor, a mammal will have to be in charge. And before that can happen, our physicians will have to recapture their belief in the substantive nature of emotional life and the determination to fight for it.

Walker Percy wrote that "modern man is estranged from being, from his own being, from the being of other creatures in the world, from transcendent being. He has lost something—what, he does not know; he only knows that he is sick unto death with the loss of it." The mysterious, absent element is a deep and abiding immersion in communal ties. In all of its varied and protean forms, love is the tether binding our whirling lives. Without that biological anchor, all of us are flung outward, singly, into the encroaching dark.

Humanity began in a precarious world where tiny bands foraged and scrimped for food by day, huddled together for warmth by night. With the advent of agriculture came mass aggregation in towns and cities. The industrial revolution took work out of the home, making the populace "a mass of undifferentiated equals, working in a factory or scattered between the factories, the mines, and the offices, bereft forever of the feeling that work was a family affair, done within the household." Economies prospered as families dissipated. In pursuit of further riches, the information age demands a more thoroughgoing surrender—less time for relationships, less time for children, more time for impersonal everything. Before our lives wither away into dust, we might ponder how much more prosperity human beings can possibly survive.

A good deal of modern American culture is an extended experiment in the effects of depriving people of what they crave most. The consequences that flow from limbic ignorance are as grim as love's victories are miraculous. The tragedy lies, as all tragedy does, in the knowledge that these sad outcomes once held the potential for greatness. What Charles Dickens wrote of the pair beneath the robe of Christmas Present is true of our society: "Where angels might have sat enthroned, devils lurked, and glared out menacing. No change, no degradation, no perversion of humanity, in any grade, through all the mysteries of wonderful creation, has monsters half so horrible and dread." Our culture might trade back these devils for the divinity that our mammalian heritage accords us, if only we had the inclination to attend to limbic imperatives.

The capacious and monocular neocortical brain tells us that ideas perpetuate civilization. The thick marble walls of libraries and museums protect our supposed bequest to future ages. How short a vision. Our children are the builders of tomorrow's world—quiet infants, clumsy toddlers, and running, squealing second graders, whose pliable neurons carry within them all humanity's hope. Their flexible brains have yet to germinate the ideas, the songs, the societies of tomorrow. They can create the next world or they can annihilate it. In either case, they will do so in our names.

Ten

THE OPEN DOOR

WHAT THE FUTURE HOLDS FOR THE MYSTERIES OF LOVE

In that abyss I saw how love held bound
Into one volume all the lives whose flight
Is scattered through the universe around;
How substance, accident, and mode unite,
Fused, so to speak, together in such wise
That this I tell is one simple light.

—Dante, from *The Divine Comedy*

The natural phenomena that first evoked mankind's passion for elucidation, definition, prediction—the passion for science—were the regular tracks of heavenly bodies against the night sky. Logical deduction led early astronomers to propose a scheme with Earth as the motionless hub of a wheeling universe. They imagined the stars and planets embedded in rotating crystal shells, thus accounting for the jeweled processions they witnessed above.

The demands of pragmatism soon impinged on the model's lovely simplicity. Increasingly careful measurements of planetary and celestial motion necessitated adding more hollow spheres to keep the system aligned with observation. Eudoxus postulated twenty-seven separate glass bowls. By the time Aristotle published his version, he needed fifty-five. Claudius Ptolemy tacked on some further awkward correctives for accuracy's sake, and that star-tracking contraption limped along for another seventeen hundred years.

The telescope's invention finally opened humanity's door to the stars and proved Earth a spinning, speeding top, caroming around the sun. Much of the motion the old astronomers had charted was the hurtling of the ground beneath their feet. The Ptolemaic machine, which had served and deceived astronomers for centuries, wheezed and collapsed.

Our age is ready to retire another host of unwieldy and outmoded contrivances: the models of the emotional mind that predated empiricism. Science is busily disassembling the brain that engenders intelligence, reason, passion, and love—the delicate structure that creates each of our selves, each of our hearts. The spectral inhabitants that we were taught to expect there—id, ego, Oedipus—are fading like summer stars before the coming dawn. The discoveries effacing these former luminaries have come from a science that explored—much as any poet must—a fantastic world to seek its principles without first assuming their continuity with the rational.

Ptolemy's geocentric map of the universe, while extravagantly incorrect, did inform sky watchers for centuries about the apparent movements of stars and planets, sun and moon. The model predicted and explained, often quite accurately. It could not explain *enough*. At the limits of its applicability, no amount of tinkering and tacking on could save it. So it is with twentieth-century models of the emotional mind. They have served and deceived faithfully, but now they creak and fray under the steady pressure of inconsistencies they cannot redress. We are not obliged to discard *all* observations made under old auspices—a core crew of findings will sail on into the future, like sailors rescued from a sinking ship, who, from their new perch on the top decks, watch an outdated vessel disappear under the waves.

Emotions reach back 100 million years, while cognition is a few hundred thousand years old at best. Despite their youth, the prominent capacities of the neocortical brain dazzled the Western world and eclipsed the mind's quieter limbic inhabitant. Because logic and deduction accomplish so plainly, they have been presumed the master keys that open all doors. The mind's early surveyors drew their plans by the light of this guiding disposition, including as weight-bearing beams their faith in a veridical reality, the supremacy of analytic thought, and the ultimate rule of rationality.

The mind's early pioneers imagined a castle in the sky that flew Reason's banner, and for too long, our society tried to live within their fanciful construction. Their efforts were commendable, even visionary, but the structure they raised proved inhospitable to the glorious illogic of human lives.

Limbic resonance, regulation, and revision define our emotional existence; they are the walls and towers of the neural edifice evolution has built for mammals to live in. Our intellect is largely blind to them. Within the heart's true edifice, those who allow themselves to be guided by Reason blunder into walls and stumble over sills. They are savants who can see too little of love to escape painful collisions with its unforgiving architecture.

Our culture teems with experts who propose to tell us how to think our way to a better future, as if that could be done. They capitalize on the ease of credibly presuming, without a pause or backward glance, that intellect is running the show. Not so. *Reason's last step*, wrote Blaise Pascal, *is recognizing that an infinity of things surpass it.* As a new millennium commences, science is beginning to approach that pinnacle of perspicacity.

Although our culture may oppose them at every turn, people can still manage to lead successful lives, if they cultivate the connections their limbic brains demand. No matter what humanity's future holds, we will never shed our heritage as neural organisms, mammals, primates. Because we are emotional beings, pain is inevitable and grief will come; because the world is neither equitable nor fair, the suffering will not be distributed evenly. A person who intuits the ways of the heart stands a better chance of living well. A society of those who do so holds a promise we can only guess at.

The adventure of seeking a theory of love is far from over. While science can afford us a closer glimpse of *this* tower or *that* soaring wall, the heart's castle still hangs high in the heavens, shrouded in scudding clouds and obscured by mist. Will science ever announce the complete revelation of all of love's secrets? Will empiricism ever trace an unbroken path from the highest stones of the heart's castle down to the bedrock of certitude?

Of course not. We demand too much if we expect single-handed empiricism to define and lay bare the human soul. Only in concert with art does science become so precise. Both are metaphors through which we strive to know the world and ourselves; both can illuminate inner and outer landscapes with a flash that inspires but whose impermanence necessitates unending rediscovery. Carl Sandburg once wrote that poetry is the opening and closing of a door, leaving those who look through to guess about what was seen during a moment. The most we can reasonably ask is that science open a door of its own from time to time, and allow us to spy for a fraction of a second the bounteous secrets inside.

Humanity awaits the revelations that may glint through that open portal. In the end, all such discovery has one purpose: to help people reach their potential for fulfillment and joy. While we cannot alter the nature of love, we can choose to defy its dictates or thrive within its walls. Those with the wisdom to do so will heed their hearts and draw strength from their relatedness, and they will raise their children to do likewise.

NOTES

Chapter I: The Heart's Castle

Page 4   *The heart has its reasons:* Pascal, 1972, p. 277.

Page 7   "Man is a credulous animal": Russell, 1950, p. 111.

Page 8   For a thorough recounting of psychiatrists' tendency to issue dogmatic and unfounded claims, see Dolnick, 1998.

Page 10   See, for instance, Steven Pinker's *How the Mind Works,* 1997. Pinker, evolutionary psychology's most articulate advocate, on friendship: "Once you have made yourself valuable to someone, the person becomes valuable to you. You value him or her because if you were ever in trouble, they would have a stake— albeit a selfish stake—in getting you out.... This runaway process is what we call friendship" (pp. 508–9). (A conception of amity undone by the mere frequency of liking a person who can do nothing to help you, and not liking many who can.) Music is "useless" (p. 528), "auditory cheesecake" (p. 534) that "communicates nothing but formless emotion" (p. 529). Religion is "a desperate measure that people resort to when the stakes are high and they have exhausted the usual techniques for the causation of success" (p. 556). "For anyone with a persistent intellectual curiosity, religious explanations are not worth knowing" (p. 560). A work of art, "having no practical function" (p. 544), exists solely to flaunt one's impressive idleness: "The very uselessness of art ... makes it all too comprehensible to economics and social psychology. What better proof that you have money to spare than your being able to spend it on doodads and stunts that don't fill the belly or keep the rain out?" (p. 522).

Page 11   Nabokov, as quoted by Appel, 1967.

Page 12   For more on the relationship between science and art, see Wilson, 1998.

Chapter 2: Kits, Cats, Sacks, and Uncertainty

Page 16   "O body swayed to music": Yeats, 1996.

Page 17   "Science does not describe and explain nature": Heisenberg, 1999, p. 81.

Page 21   Evolution as continuous versus abrupt change: Eldredge and Gould, 1972.

Page 21   Triune brain model: MacLean, 1973; 1985; 1990.

Page 22   We use "reptilian brain" to denote the structures that MacLean calls the "R complex" and the "protoreptilian formation."

Page 24   Derivation of the term "limbic": Broca, as quoted by Schiller, 1992.

Page 24   We employ the term "limbic brain" in place of MacLean's synonymous appellation, "paleomammalian brain."

Page 27   In place of MacLean's "neomammalian brain," we substitute the phrase "neocortical brain."

Page 29   Regaining command of muscles after a stroke: Liepert et al., 1998.

Page 29   Readiness wave before conscious awareness of decision: Libet et al., 1983. Dennett and Kinsbourne (1995) point out that the temporal order in which events appear to occur to one subjective conscious mind is itself fraught with potential unreliability.

Page 29   Always a more beautiful answer: E. E. Cummings, 1994.

Page 30   Neocortex as a means to ballistic motor movements: Calvin, 1994; Calvin, 1996.

Page 32   Radiographic dyes selectively staining the limbic brain: MacLean, 1990.

Page 32   A monoclonal antibody to limbic system neurons: Levitt, 1984.

Page 32   Selective destruction of limbic tissue: Peredery et al., 1992.

Page 32   Neocortical lesions spare maternal abilities while limbic damage destroys them: MacLean, 1990.

Page 32   Limbic lesions in monkeys obliterating social awareness: Ward, 1948.

Page 32   Limbectomized hamsters suffering loss of social sensation: MacLean, 1990.

Page 32   Einstein on the limitations of intellect: Einstein, 1995, p. 260.

Page 33   "From modern neuroanatomy, it is apparent": Tucker and Luu, 1998.

Page 33   "We say, 'I will,' and 'I will not,' and imagine ourselves": Wolfe, 1994, p. 110.

Chapter 3: Archimedes' Principle

Page 37   Biological purpose of facial expressions of emotion: Darwin, 1998.

Page 39   Universality of facial expression: Izard, 1971; Ekman et al., 1972.

Page 39   Facial expressions in New Guinea: Ekman, 1973; Ekman, 1984.

Page 40   Facial expressions in nonhuman mammals: Chevalier-Skolnikoff, 1973; Redican, 1982.

Page 40   For one scientific endorsement of animal feelings, see Panksepp, 1994.

Page 40   Mark Derr on animal consciousness, intelligence, volition, and feelings: Derr, 1997.
Page 43   The brevity of facial expressions: Ekman, 1992.
Page 44   Definition of mood: Ekman, 1992.
Page 47   Hardwired temperament: Emerson, 1979.
Page 48   Centers in the reptilian brain (brain stem) controlling temperament: Cloninger et al., 1993.
Page 48   Anxiety controlled by the raphe nucleus: Cloninger et al., 1993.
Page 49   Criminality as partly inherited temperament: Brennan et al., 1997; Raine et al., 1994.
Page 50   Autonomic nervous system too far from the head: Semrad, as quoted by Rako and Mazer, 1980.
Page 58   Right temporal neocortex processing emotional intonation of speech: Ross, 1981.
Page 60   Babies' propensity for looking at faces: Johnston et al., 1991.
Page 61   Infants' ability to distinguish among facial expressions: Field et al., 1982.
Page 61   Visual cliff: Emde, 1983.
Page 62   A mother's freezing her expression upsets her baby: Brazelton et al., 1975.
Page 62   Videotaping the interaction between mothers and babies: Trevarthen, 1993.

Chapter 4: A Fiercer Sea

Page 66   "But if thou, jealous, dost return to pry": *Romeo and Juliet*, Act V, Scene 3.
Page 67   Lorenz's upbringing: Lorenz, 1973.
Page 68   Lorenz and studies of avian mother-offspring bonding: Lorenz, 1973.
Page 68   Lambs bonding to television sets: Cairns, 1966.
Page 68   Guinea pigs bonding to wooden blocks: Shipley, 1963.
Page 68   Monkeys bonding to cylinders of wire: Harlow, 1958.
Page 69   Frederick's experiment: Coulton, 1906, pp. 242–43.
Page 70   The medical morbidity of the sterile nurseries: Spitz, 1945.
Page 70   Attachment theory: Bowlby, 1983, 1986.
Page 71   Anna Freud rebukes Bowlby: Freud, A., 1960.
Page 71   Winnicott's revulsion: Winnicott, as quoted by Bretherton, 1991.
Page 71   Bowlby's therapist stands up and denounces him: Karen, 1994. (Karen's book—*Becoming Attached: First Relationships and How They Shape Our Capacity to Love*—is one of the very few books that we routinely recommend to people wishing to understand more about how relationships function. It has informed the material in this chapter and is well worth reading.)

Page 71   Behaviorist Watson's advice to parents: Watson, 1928. Nor was
               this the most extreme of Watson's views. "It is a serious ques-
               tion in my mind," he wrote, "whether there should be individ-
               ual homes for children—or even whether children should know
               their own parents. There are undoubtedly much more scientific
               ways of bringing up children." Mercifully, not everyone fol-
               lowed Watson's advice. After one of Watson's lectures, a woman
               came up to him and said, "Thank God my children are
               grown . . . and that I had a chance to enjoy them before I met
               you."

Page 72   Wire-monkey mother studies: Harlow, 1958.

Page 72   Unique sound signature of an infant's hunger cry: Zeskind et
               al., 1993.

Page 74   Mothering style linked to a child's developmental outcome:
               Ainsworth et al., 1978.

Page 74   School age outcome of secure and insecure attachment:
               Ainsworth et al., 1978; Bretherton, 1985.

Page 75   Adolescent outcome of secure and insecure attachment: Karen,
               1994.

Page 76   Universal mammalian separation reaction: a nice summary can
               be found in Hofer, 1984.

Page 76   Physiology of the protest phase: Hofer, 1984.

Page 77   Elevations of catecholamines and cortisol in the protest phase
               of separation: Gunnar et al., 1981; Breese et al., 1977.

Page 78   Sixfold increase in cortisol after thirty minutes of separation:
               Shapiro and Insel, 1990.

Page 78   Physiology of the despair phase: Hofer, 1984.

Page 79   Immune regulation in despair: Hofer, 1984.

Page 80   Social isolation and mortality after heart attack: Berkman et al.,
               1992.

Page 80   Group psychotherapy and survival after metastatic breast can-
               cer: Spiegel et al., 1989.

Page 80   Social support and survival rates in leukemia: Colon et al.,
               1991.

Page 80   Solitary people have vastly increased rates of premature death
               from all causes: Ornish, 1998.

Page 81   Hofer's finding that a nonmaternal heat source does not main-
               tain rat pup body temperature: Hofer, 1975; Hofer, 1995.

Page 82   Single maternal attributes regulating specific aspects of infant
               physiology: Hofer, 1984; Hofer, 1987.

Page 83   Inability of mammals to maintain solo neurophysiologic stabil-
               ity: Hofer, 1984; Field, 1985; Kraemer, 1985; Reite and Capi-
               tanio, 1985; Kraemer, 1992.

Page 85   Influence of the breathing bear on premature infants: Thoman
            et al., 1991; Ingersoll and Thoman, 1994.
Page 86   "I went to the woods because I wished to live deliberately":
            Thoreau, 1960, p. 66.
Page 87   "And say'st thou yet that exile": *Romeo and Juliet*, Act III, Scene 3.
Page 87   For the simplest demonstration that lack of nurturance causes
            structural brain abnormalities, see Martin et al., 1991, and
            Floeter and Greenough, 1979; Lewis et al., 1990.
Page 88   The isolation syndrome: Kraemer, 1992.
Page 89   Hubel and Wiesel, 1970.
Page 90   Alterations in the adult brains of monkeys raised by distracted
            mothers: Andrews and Rosenblum, 1994; Rosenblum and An-
            drews, 1994; Rosenblum et al., 1994; Coplan et al., 1998.
Page 94   Sydenham in praise of opiates: as quoted by Smith, 1995.
Page 94   Opiate receptors in limbic areas: Wise and Herkenham, 1982.
Page 94   Small doses of opiates erase protest in isolated puppies:
            Panksepp et al., 1985.
Page 94   "A twinging ache of grief rose up": Book 4 of Homer's *The
            Odyssey*, trans. by R. Fitzgerald, 1963. (We first encountered
            these lines in Panksepp, 1992.)
Page 96   Oxytocin and the brains of prairie dogs: Insel, 1992.
Page 97   Oxytocin levels surge around birth: Nissen et al., 1995.
Page 98   Pet ownership and improved survival in cardiac patients: Fried-
            mann and Thomas, 1995; Friedmann et al., 1980.
Page 98   "Although we are by all odds the most social of all social ani-
            mals": Thomas, 1995, p. 14.

Chapter 5: Gravity's Incarnation

Page 100   "Memory collects the countless phenomena": Hering, 1911.
            (We first encountered this lovely description in D. L. Schacter,
            1990.)
Page 101   "In literature, as in love": Maurois, 1963.
Page 101   "From the repressed memory traces, it can be verified": Freud,
            1938, p. 174.
Page 105   Effects of hippocampal damage on memory: Squire et al.,
            1993.
Page 107   Predicting the weather: Knowlton et al., 1996.
Page 109   Cognitive efforts impair performance in implicit tasks: Reber,
            1976. Pretask explanations fail to improve performance on im-
            plicit tasks: Berry and Broadbent, 1984.
Page 109   Artificial grammar implicit task: Knowlton et al., 1992.
Page 110   Implicit gambling task: Bechara et al., 1997.

Page 112   Implicit nature of phonological and grammatical rules: Pinker, 1995, pp. 174, 273.

Page 114   Letter from Freud to Fliess in which the former claims he was able to "hear again the words that were exchanged between two adults": Freud, S., as quoted by Borch-Jacobsen, 1998, p. 50.

Page 114   Babies' memory for mothers' voice and face shortly after birth: DeCasper and Fifer, 1980.

Page 114   Effect of mother's language on infants: Moon et al., 1993; Fifer and Moon, 1994.

Page 115   The adventure of patient Boswell: Tranel and Damasio, 1993.

Page 115   The devil's finest trick is convincing the world he doesn't exist: Baudelaire, 1998. His character says, "*Mes chers frères, n'oubliez jamais . . . que la plus belle des ruses du diable est de vous persuader qu'il n'existe pas!*" After disappearing for a century, the line popped up in the middle of the film *The Usual Suspects.*

Page 119   "All things appear to us as they appear": Eco, 1995.

Chapter 6: A Bend in the Road

Page 122   Computer program that diagnoses heart attack: Baxt, 1991. For physicians, the sensitivity (how often their prediction of heart attack was borne out) was 78 percent, and their specificity (how often their exclusion of heart attack was correct) was 85 percent. The computer had a sensitivity of 97 percent and a specificity of 96 percent.

Page 122   Computers excelling at various diagnostic tasks: Baxt, 1995.

Page 123   The role of neural networks in mutual fund forecasting: Ridley, 1993.

Page 123   Impossibility of determining how a neural network makes the predictions it does: Robinson, 1992.

Page 125   Neural network modeling: Rumelhart and McClelland, 1986.

Page 127   "This storage scheme": Hebb, 1949.

Page 127   Experimental validation of Hebb's hypothesis: Ahissar et al., 1992.

Page 131   Abused children exhibit exaggerated brain waves upon seeing an angry face: Pollak, 1999.

Page 134   Fallibility of eyewitness testimony: Loftus, 1993.

Page 135   Memory for the *Challenger* explosion: Neisser and Harsch, 1992.

Page 138   The Kanisza triangle: While we have described the process whereby experience creates Attractors, the same basic neural structure—a collection of powerful links that channels perception and determines experience—can as easily be part of the brain's genetically endowed wiring scheme.

Page 139   Learned Attractor for H's and A's: adapted from Selfridge, 1955.

Page 141   "In scientific work, we find": Tucker and Luu, 1998.

Page 142   Dr. Remen's story: Remen and Ornish, 1997.

Page 143   "your homecoming will be my homecoming": Poem 40 in "73 Poems," E. E. Cummings, 1994.

Chapter 7: The Book of Life

Page 145   "From a drop of water": "A Study in Scarlet," by Arthur Conan Doyle.

Page 145   "compelled to reason backward from effects to causes": "The Adventure of the Cardboard Box," by Arthur Conan Doyle.

Page 146   The logical impossibility of the Holmesian method: See, for example, Truzzi, 1983, who concludes that "[t]he simple fact is that the vast majority of Holmes's inferences just do not stand up to logical examination. He concludes correctly simply be- cause the author of the stories allows it so."

Page 151   "My heart, still full of her": From "Souvenir," by Alfred de Musset.

Page 152   Anxiety-prone rodents: Eley and Plomin, 1997.

Page 153   Nurturant rearing can override anxious temperament in mon- keys: Suomi, 1991.

Page 153   "In faith, I do not love thee": Shakespeare, Sonnet 141.

Page 157   Increased rate of depression in mothers of children with SIDS: Mitchell et al., 1992.

Page 157   Heart rhythm stability and attachment security: Izard et al., 1991.

Chapter 8: Between Stone and Sky

Page 168   Psychotherapy alters the brain: Baxter et al., 1992, found that behavior therapy and Prozac produced similar changes in the basal ganglia of patients suffering from obsessive-compulsive disorder as observed by PET (positron emission tomography) scanning. Viinamki et al., 1998, demonstrated via SPECT (sin- gle positron emission computed tomography) scanning that psychotherapy changed serotonin uptake in the basal forebrain of a patient with personality disorder and depression. At the 1999 meeting of the American Psychiatric Association, Dr. Stephen D. Martin reported that SPECT scanning shows that interpersonal psychotherapy causes changes in the basal ganglia and cingulate gyrus of patients with depression (as described by Zwillich, 1999).

Page 168   "Where id was": Freud, 1953, Volume XXII, p. 80.

Page 172   Inability of depressed people to recognize facial expressions of
            emotion: Rubinow and Post, 1992. See also Bench et al., 1993,
            and Mayburg et al., 1994, who both found substantial de-
            creases in limbic brain activity in depressed patients.

Page 174   Mistaken assumptions about placebos and the effectiveness of
            medications: See, for example, Horgan, 1999. His account of
            the placebo response manages to cram conceptual confusion
            and factual errors into just over five hundred words, which may
            be a record for achieving a local concentration of misstatement.

Page 176   Mechanisms of antidepressant efficacy are mysterious: Most
            people have heard that antidepressants work by "raising sero-
            tonin," a descriptive account but one devoid of explanatory
            power. (One might similarly observe that in open-heart surgery,
            the doctor makes an incision and hours later the patient is
            cured.) The inner workings of antidepressant efficacy, what al-
            tering serotonin receptor binding patterns does in the brain
            that culminates days or weeks later in mood change, remain to-
            tally obscure.

Page 176   Seven plus three equals ten: Augustine, as quoted by Park,
            1988.

Page 179   Teaching Japanese speakers to distinguish between "r" and "l":
            Blakeslee, 1999.

Page 182   "We should always remember": Nabokov, 1980.

Page 183   Art "as a more or less shocking surprise": Nabokov, 1955.

Page 184   Freud on the emotional coldness he thought requisite for psy-
            chotherapy: Freud, 1953, Volume XII, pp. 115, 118.

Page 187   Bok's words: "If you think education is expensive, try igno-
            rance."

Page 188   "Who needs psychoanalysis": Michael Freeman, as quoted in
            Hymowitz and Pollac, 1995.

Page 190   Only human love keeps surgery from being an act between two
            madmen: Selzer, 1982, p. 106.

Chapter 9: A Walk in the Shadows

Page 193   Spock on children in the parents' bed: Spock, 1945, p. 169.

Page 193   Ferber on the multifarious hazards of cosleeping: Ferber, 1986.

Page 194   Robert Wright refutes Ferber: Wright, 1997.

Page 195   Cultural variability in SIDS rates: McKenna, 1996.

Page 195   "The temporal unfolding of particular sleep stages": McKenna
            et al., 1990.

Page 195   Correlation of SIDS rates and intolerance of cosleeping:
            McKenna et al., 1993.

Page 199   Internet use causes depression: Kraut et al., 1998.

Page 199   "We were surprised to find that": Carnegie Mellon University press release, September 1, 1998.

Page 199   An emotional void's lethal impact on babies: Spitz, 1945.

Page 199   Neglect produces children with smaller head circumferences and brain shrinkage visible on scanning: Perry and Pollard, 1997.

Page 199   Maternal depression produces cognitive deficits in children: Murray et al., 1991; Murray, 1992.

Page 199   Attachment research showing that responsive parenting confers long-lasting personality strengths: For a lucid summary, see Karen, 1994.

Page 199   The neural devastation following isolation of young monkeys: Kraemer, 1985; Kraemer, 1992; Kraemer and Clarke, 1990; Kraemer and Clarke, 1996.

Page 199   Producing permanent changes in a monkey's brain by subjecting his mother to conditions that create an emotional state of uncertainty: Andrews and Rosenblum, 1994; Rosenblum and Andrews, 1994; Rosenblum et al., 1994; Coplan et al., 1998.

Page 199   Degree of maternal nurturance related to the stress resilience of adult rats: Liu et al., 1997; Meaney et al., 1991.

Page 202   Developmental psychology researchers refusing to voice an opinion on the impact of day care: Karen, 1994, p. 230.

Page 204   Impact of divorce on children: Wallerstein and Kelly, 1980.

Page 209   Bertrand Russell on love: Russell, 1998.

Page 211   Neglected children become sensitized to stress: Francis et al., 1996.

Page 211   Suicide rates in adolescents: Singh and Yu, 1996.

Page 213   Studies demonstrating that close family relationships produce a substantial decrease in rates of adolescent substance abuse: Just to cite a few, see Bell and Champion, 1979; Nicholi, 1983; Jurich et al., 1985; Cadoret et al., 1986; Reynolds and Rob, 1988; Stoker and Swadi, 1990; Nurco et al., 1996; Risser et al., 1996; Nurco et al., 1998.

Page 214   "a clarification of life": Frost, 1995, p. 777.

Page 215   "When you love, you wish to do things for": Hemingway, 1995, p. 72.

Page 217   "I said, 'Mr. Brown, do you mean to tell me' ": Testimony of Charles Roemer about the meeting, in 1942 or 1943, of Unarco officials with Johns Manville president Lewis Brown and his brother Vandiver Brown. Deposition taken April 25, 1984, *Johns-Manville Corp et al. v. the United States of America*, U.S. Claims Court Civ. No. 465-83C. Castleman and Berger, 1996. (Our deep appreciation to Roger G. Worthington for supplying this citation.) See also Brodeur, 1985.

Page 218   Patterns of correspondence between aggression and levels of serotonin, norepinephrine, and dopamine: Kraemer and Clarke, 1990; 1996.

Page 218   Isolation-reared monkeys do not "conform to the usual social rules": Kraemer and Clarke, 1996.

Page 218   "What do I care?": Damon, 1999.

Page 221   "I wouldn't demand a lot of my doctor's time": Broyard, 1990.

Page 222   Teaching acting to medical students: Finestone and Conter, 1994.

Page 224   Managed care physician confesses to causing a patient's death by denial of care: Carlsen, 1997.

Page 224   New York managed care organization uses death rate data to funnel patients into the worst and cheapest facility: National Public Radio broadcast of "Marketplace," December 3, 1997.

Page 224   "modern man is estranged from being, from his own being": Percy, 1992.

Page 225   Impact of industrial revolution on the family: Laslett, 1984, p. 18.

Chapter 10: The Open Door

Page 229   *Reason's last step:* Pascal, 1972, p. 267.

BIBLIOGRAPHY

Ahissar, E., E. Vaadia, M. Ahissar, H. Bergman, A. Arieli, and M. Abeles. 1992. "Dependence of cortical plasticity on correlated activity of single neurons and on behavioral context." *Science*, 257(5075):1412–5.

Ainsworth, M., M. Blehar, E. Waters, and S. Wall. 1978. *Patterns of Attachment: A Psychological Study of the Strange Situation.* Mahwah, New Jersey: Erlbaum & Associates.

Andrews, M. W., and L. A. Rosenblum. 1994. "The development of affiliative and agonistic social patterns in differentially reared monkeys." *Child Development*, 65(5):1398–1404.

Appel, A. "An Interview with Nabokov." 1967. *Wisconsin Studies in Contemporary Literature* (8). As quoted by L. Shlain in *Art & Physics: Parallel Visions in Space, Time, and Light.* 1991. New York: William Morrow.

Baudelaire, C. "Le Joueur Généreux." 1998. In *Le Spleen de Paris.* Paris: Librairie Générale Française.

Baxt, W. G. 1991. "Use of an artificial neural network for the diagnosis of myocardial infarction." *Annals of Internal Medicine*, 115(11):843–8.

Baxt, W. G. 1995. "Application of artificial neural networks to clinical medicine." *The Lancet*, 346(8983):1135–8.

Baxter, L. R., J. M. Schwartz, K. S. Bergman, M. P. Szuba, B. H. Guze, J. C. Mazziotta, A. Alazraki, C. E. Selin, H. K. Ferng, and P. Munford. 1992. "Caudate glucose metabolic rate changes with both drug and behavior therapy for obsessive-compulsive disorder." *Archives of General Psychiatry*, 49(9):681–9.

Bechara, A., H. Damasio, D. Tranel, and A. R. Damasio. 1997. "Deciding advantageously before knowing the advantageous strategy." *Science*, 275(5304):1293–5.

Bell, D. and R. Champion. 1979. "Deviancy, delinquency and drug use." *British Journal of Psychiatry*, 134: 269–76.

Bench, C., K. Friston, R. Brown, L. Scott, R. Frackowiak, and R. Dolan. 1993. "The anatomy of melancholia—focal abnormalities of cerebral blood flow in major depression." *Psychological Medicine*, 22: 607–15.

Berkman, L. F., L. Leo-Summers, and R. I. Horwitz. 1992. "Emotional support and survival after myocardial infarction. A prospective, population-based study of the elderly." *Annals of Internal Medicine*, 117(12):1003–9.

Berry, D. C., and D. E. Broadbent. 1984. "On the relationship between task performance and associated verbalizable knowledge." *The Quarterly Journal of Experimental Psychology,* 36A: 209–231.

Blakeslee, S. "Old Brains Can Learn New Language Tricks." 1999. *The New York Times,* April 20.

Borch-Jacobsen, M. 1998. "Self-Seduced." In F. Crews (ed.), *Unauthorized Freud: Doubters Confront a Legend.* New York: Viking.

Bowlby, J. 1983. *Attachment & Loss,* Volume I: Attachment. New York: Basic Books.

Bowlby, J. 1986. *Attachment & Loss,* Volume II: Separation. New York: Basic Books.

Bowlby, J. 1986. *Attachment & Loss,* Volume III: Loss. New York: Basic Books.

Brazelton, T. B., E. Tronick, L. Adamson, H. Als, and S. Wise. 1975. "Early mother-infant reciprocity." *Ciba Foundation Symposium,* 33: 137–54.

Breese, G. R., R. D. Smith, R. A. Mueller, J. L. Howard, A. J. Prange, M. A. Lipton, L. D. Young, W. McKinney, and J. K. Lewis. 1977. "Induction of adrenal catecholamine synthesizing enzymes following mother-infant separation." *Nature New Biology,* 246: 94–6.

Brennan, P. A., A. Raine, F. Schulsinger, L. Kirkegaard-Sorenson, J. Knop, B. Hutchings, R. Rosenberg, and S. A. Mednick. 1997. "Psychophysiological protective factors for male subjects at high risk for criminal behavior." *American Journal of Psychiatry,* 154(6):53–5.

Bretherton, I. 1985. "Attachment Theory: Retrospect and Prospect." In I. Bretherton and E. Waters (eds.), *Growing Points of Attachment Theory and Research (Monographs of the Society for Research in Child Development).* Chicago: University of Chicago Press.

Bretherton, I. 1991. "The roots and growing points of attachment theory." In C. M. Parkes, J. Stevenson-Hinde, and P. Marris (eds.), *Attachment Across the Life-Cycle.* New York: Tavistock/Routledge.

Brodeur, P. 1985. *Outrageous Misconduct: The Asbestos Industry on Trial.* New York: Pantheon.

Broyard, A. 1990. "Doctor Talk to Me." *The New York Times,* August 26.

Cadoret, R. J., E. Troughton, T. W. O'Gorman, and E. Heywood. 1986. "An adoption study of genetic and environmental factors in drug abuse." *Archives of General Psychiatry,* 43(12):1131–6.

Cairns, R. B. 1966. "Attachment behavior of mammals." *Psychological Review,* 73: 409–26.

Calvin, W. H. 1996. *How Brains Think: Evolving Intelligence, Then and Now.* New York: HarperCollins.

Calvin, W. H. 1994. "The emergence of intelligence." *Scientific American,* 271(4):100–7.

Carlsen, W. 1997. "Doctor to Confess Role in Man's Death: she says HMO rewarded her for saving $500,000." *San Francisco Chronicle,* April 15, p. A13.

Castleman, B., and S. Berger. 1996. *Asbestos: Medical and Legal Aspects.* Gaithersburg, Maryland: Aspen Publishers.

Cave, C. B., and L. R. Squire. 1992. "Intact and long-lasting repetition priming in amnesia." *Journal of Experimental Psychology. Learning, Memory, and Cognition,* 18(3):509–20.

Chevalier-Skolnikoff, S. 1973. "Facial expressions of emotion in nonhuman primates." In P. Ekman (ed.), *Darwin and Facial Expression: A Century of Research in Review.* New York: Academic Press.

Cloninger, C. R. 1986. "A unified biosocial theory of personality and its role in the development of anxiety states." *Psychiatric Developments,* 4(3):167–226.

Cloninger, C. R. 1987. "A systematic method for clinical description and classification of personality variants." *Archives of General Psychiatry,* 44: 573–87.

Cloninger, C. R., D. M. Svrakic, and T. R. Przybeck. 1993. "A psychobiological model of temperament and character." *Archives of General Psychiatry,* 50(12):975–90.

Colon, A., A. Callies, M. Popkin, and P. McGlave. 1991. "Depressed mood and other variables related to bone marrow transplantation survival in acute leukemia." *Psychosomatics,* 32: 420–5.

Coplan, J. D., R. C. Trost, M. J. Owens, T. B. Cooper, J. M. Gorman, C. B. Nemeroff, and L. A. Rosenblum. 1998. "Cerebrospinal fluid concentrations of somatostatin and biogenic amines in grown primates reared by mothers exposed to manipulated foraging conditions." *Archives of General Psychiatry,* 55(5):473–7.

Coulton, G. G. 1906. *St. Francis to Dante.* London: David Nutt.

cummings, e. e. 1994. *E. E. Cummings: Complete Poems, 1904–1962.* G. Firmage (ed.). New York: Liveright.

Damon, W. 1999. "The Moral Development of Children." *Scientific American,* 281(2):72–8.

Darwin, C. 1998. *The Expressions of the Emotions in Man and Animals.* Third Edition. P. Ekman (ed.). New York: Oxford University Press.

DeCasper, A. J., and W. P. Fifer. 1980. "Of human bonding: newborns prefer their mothers' voices." *Science,* 208(4448):1174–6.

Dennett, D. C., and M. Kinsbourne. 1995. "Time and the observer: the where and when of consciousness in the brain." *Behavioral and Brain Sciences*, 15(2):183–247.

Derr, M. 1997. "Puppy Love." *The New York Times Book Review*, October 5.

Dolnick, E. 1998. *Madness on the Couch: Blaming the Victim in the Heyday of Psychoanalysis.* New York: Simon & Schuster.

Eco, U. 1995. *The Island of the Day Before.* New York: Harcourt Brace and Company.

Einstein, A. 1995. "The Goal of Human Existence." *Out of My Later Years.* New York: Carol Publishing Group.

Ekman, P. 1973. "Cross-Cultural Studies of Facial Expression." In P. Ekman (ed.), *Darwin and Facial Expression: A Century of Research in Review.* New York: Academic Press.

Ekman, P. 1984. "Expression and the Nature of Emotion." In K. Scherer and P. Ekman (eds.), *Approaches to Emotion.* Hillsdale, New Jersey: Lawrence Erlbaum.

Ekman, P. 1992. "An argument for basic emotions." *Cognition and Emotion*, 6(3/4): 169–200.

Ekman, P., W. V. Friesen, and P. Ellsworth. 1972. *Emotion in the Human Face: Guidelines for Research and an Integration of Findings.* New York: Pergamon Press.

Eldredge, N., and S. J. Gould. 1972. "Punctuated equilibria: an alternative to phyletic gradualism." In T.J.M. Schopf (ed.), *Models in Paleobiology.* San Francisco: Freeman, Cooper.

Eley, T. C., and R. Plomin. 1997. "Genetic analyses of emotionality." *Current Opinion in Neurobiology*, 7(2):279–84.

Emde, R. N. 1983. "The prerepresentational self and its affective core." *The Psychoanalytic Study of the Child*, 38: 165–92.

Emerson, R. W. 1979. "Experience." *The Essays of Ralph Waldo Emerson.* Cambridge, Massachusetts: Harvard University Press.

Erickson, M. T. 1993. "Rethinking Oedipus: an evolutionary perspective of incest avoidance." *American Journal of Psychiatry*, 150: 3.

Ferber, R. 1986. *Solve Your Child's Sleep Problems.* New York: Simon & Schuster.

Field, T. 1985. "Attachment as psychobiological attunement: being on the same wavelength." In M. Reite and T. Field (eds.), *The Psychobiology of Attachment and Separation.* New York: Academic Press.

Field, T. M., R. Woodson, R. Greenberg, and D. Cohen. 1982. "Discrimination and imitation of facial expressions by neonates." *Science*, 218(8):179–81.

Fifer, W. P., and C. M. Moon. 1994. "The role of mother's voice in the organization of brain function in the newborn." *Acta Paediatrica Supplement*, 397: 86–93.

Finestone, H. M., and D. B. Conter. 1994. "Acting in medical practice." *The Lancet*, 344: 801–2.

Floeter, M. K., and W. T. Greenough. 1979. "Cerebellar plasticity: modification of Purkinje cell structure by differential rearing in monkeys." *Science*, 206(4415):227–9.

Francis, D., J. Diorio, P. LaPlante, S. Weaver, J. R. Seckl, and M. J. Meaney. 1996. "The role of early environmental events in regulating neuroendocrine development: moms, pups, stress, and glucocorticoid receptors." *Annals of the New York Academy of Sciences*, 794: 136–52.

Freud, A. 1960. *The Psychoanalytic Study of the Child*, 15: 53–62.

Freud, S. 1938. "Psychopathology of Everyday Life." In A. Brill (ed.), *The Basic Writings of Sigmund Freud*. New York: Random House.

Freud, S. 1953. In J. Strachey (ed.), *The Standard Edition of the Complete Psychological Works of Sigmund Freud*. London: Hogarth Press.

Friedmann, E., and S. A. Thomas. 1995. "Pet ownership, social support, and one-year survival after acute myocardial infarction in the Cardiac Arrhythmia Suppression Trial (CAST)." *American Journal of Cardiology*, 76: 1213–7.

Friedmann, E., A. H. Katcher, J. J. Lynch, and S. A. Thomas. 1980. "Animal companions and one-year survival of patients after discharge from a coronary care unit." *Public Health Reports*, 95(4):307–12.

Frost, R. 1995. *Collected Poems, Prose, and Plays*. New York: The Library of America.

Gavish, L., J. E. Hofmann, and L. L. Getz. 1984. "Sibling recognition in the prairie vole, Microtus ochrogaster." *Animal Behavior*, 23: 362–6.

Gunnar, M. R., C. A. Gonzalez, B. L. Goodlin, and S. Levine. 1981. "Behavioral and pituitary-adrenal responses during a prolonged separation period in infant rhesus macaques." *Psychoneuroendocrinology*, 6(1):65–75.

Harlow, H. F. 1958. "The nature of love." *American Psychologist*, 13: 673–85.

Hebb, D. O. 1949. *The Organisation of Behavior: A Neuropsychological Theory*. New York: Wiley.

Heisenberg, W. 1999. *Physics and Philosophy: The Revolution in Modern Science*. New York: Prometheus Books.

Hemingway, E. 1995. *A Farewell to Arms*. New York: Simon & Schuster.

Hering, E. 1911. "Memory as a Universal Function of Organized Matter." In S. Butler (ed.), *Unconscious Memory*. New York: E. P. Dutton.

Hinde, R. A., and L. McGinnis. 1977. "Some factors influencing the effects of temporary mother-infant separation: some experiments with rhesus monkeys." *Psychological Medicine*, 7: 197–212.

Hofer, M. A. 1975. "Survival and recovery of physiologic functions after early maternal separation in rats." *Physiology and Behavior*, 15(5):475–80.

Hofer, M. A. 1984. "Relationships as regulators: a psychobiologic perspective on bereavement." *Psychosomatic Medicine*, 46(3):183–97.

Hofer, M. A. 1987. "Early social relationships: a psychobiologist's view." *Child Development*, 58(3):633–47.

Hofer, M. A. 1994. "Early relationships as regulators of infant physiology and behavior." *Acta Paediatrica Supplement*, 387: 9–18.

Hofer, M. A. 1994. "Hidden regulators in attachment, separation, and loss." *Monographs of the Society for Research in Child Development*, 59(2–3):192–207.

Hofer, M. A. 1995. "Hidden regulators: implications for a new understanding of attachment, separation, and loss." In S. Goldberg, R. Muir, and J. Kerr (eds.), *Attachment Theory: Social, Developmental, and Clinical Perspectives*. Hillsdale, New Jersey: Analytic Press.

Hofer, M. A. 1996. "On the nature and consequences of early loss." *Psychosomatic Medicine*, 58: 570–81.

Homer. 1961. *The Odyssey*, R. Fitzgerald (trans.). New York: Doubleday.

Horgan, J. 1999. "Placebo Nation." *The New York Times*, March 21.

Hubel, D. H., and T. N. Wiesel. 1970. "The period of susceptibility to the physiological effects of unilateral eye closure in kittens." *Journal of Physiology*, 206(2):419–36.

Hymowitz, C., and E. J. Pollak. 1995. "Psychobattle: cost-cutting firms monitor couch time as therapists fret." *The Wall Street Journal*, July 13.

Ikemoto, S., and J. Panksepp. 1992. "The effects of early social isolation on the motivation for social play in juvenile rats." *Developmental Psychobiology*, 25(4):261–74.

Ingersoll, E. W., and E. B. Thoman. 1994. "The breathing bear: effects on respiration in premature infants." *Physiology and Behavior*, 56(5):855–9.

Insel, T. R. 1992. "Oxytocin: a neuropeptide for affiliation: evidence from behavioral, receptor, autoradiographic, and comparative studies." *Psychoneuroendocrinology*, 17(1):3–35.

Izard, C. E. 1971. *The Face of Emotion.* New York: Appleton-Century-Crofts.

Izard, C. E., S. W. Porges, R. F. Simons, O. M. Haynes, C. Hyde, M. Parisi, and B. Cohen. 1991. "Infant cardiac activity: developmental changes and relations with attachment." *Developmental Psychology,* 27(3):32–39.

Johnston, M. H., S. Dziurawiec, H. D. Ellis, and J. Morton. 1991. "Newborns' preferential tracking of face-like stimuli and its subsequent decline." *Cognition,* 40: 1–21.

Jurich, A. P., C. J. Polson, J. A. Jurich, and R. A. Bates. 1985. "Family factors in the lives of drug users and abusers." *Adolescence,* 20(77):143–59.

Karen, R. 1994. *Becoming Attached: First Relationships and How They Shape Our Capacity to Love.* New York: Oxford University Press.

Kihlstrom, J., T. Barnhardt, and D. Tataryn. 1992. "The psychological unconscious: found, lost, and regained." *American Psychologist,* 47(6):788–91.

Knowlton, B. J., J. A. Mangels, and L. R. Squire. 1996. "A neostriatal habit learning system in humans." *Science,* 273: 1399–1402.

Knowlton, B. J., S. J. Ramus, and L. R. Squire. 1992. "Intact artificial grammar learning in amnesia: dissociation of classification learning and explicit memory for specific instances." *Psychological Science,* 3(3):172–9.

Kraemer, G. W. 1985. "Effects of differences in early social experience on primate neurobiological-behavioral development." In M. Reite and T. Field (eds.), *The Psychobiology of Attachment and Separation.* New York: Academic Press.

Kraemer, G. W. 1992. "A psychobiological theory of attachment." *Behavioral and Brain Sciences,* 15: 493–541.

Kraemer, G. W., and A. S. Clarke. 1990. "The behavioral neurobiology of self-injurious behavior in rhesus monkeys." *Progress in Neuro-Psychopharmacology and Biological Psychiatry,* 14: S141–S168.

Kraemer, G. W., and A. S. Clarke. 1996. "Social attachment, brain function, and aggression." *Annals of the New York Academy of Sciences,* 794: 121–35.

Kraemer, G. W., M. H. Ebert, D. E. Schmidt, and W. T. McKinney. 1989. "A longitudinal study of the effect of different social rearing conditions on cerebrospinal fluid norepinephrine and biogenic amine metabolites in rhesus monkeys." *Neuropsychopharmacology,* 2(3):175–89.

Kraut, R., M. Patterson, V. Lundmark, S. Kiesler, T. Mukopadhyay, and W. Scherlis. 1998. "Internet paradox. A social technology that reduces social involvement and psychological well-being?" *American Psychologist,* 53(9):1017–31.

Laslett, P. 1984. *The World We Have Lost: Further Explored.* New York: Charles Scribner's Sons.

Levin, H. S. 1989. "Memory deficit after closed head injury." *Journal of Clinical and Experimental Neuropsychology,* 12: 129–153.

Levitt, P. 1984. "A monoclonal antibody to limbic system neurons." *Science,* 223(4633):299–301.

Lewis, M. H., J. P. Gluck, A. J. Beauchamp, M. F. Keresztury, and R. B. Mailman. 1990. "Long-term effects of early social isolation in Macaca mulatta: changes in dopamine receptor function following apomorphine challenge." *Brain Research,* 513(1):67–73.

Libet, B., C. A. Gleason, W. E. Wright, and D. K. Pearl. 1983. "Time of conscious intention to act in relation to onset of cerebral activities (readiness-potential), the unconscious initiation of a freely voluntary act." *Brain,* 106: 623–42.

Liepert, J., W. H. Miltner, H. Bauder, M. Sommer, C. Dettmers, E. Taub, and C. Weiller. 1998. "Motor cortex plasticity during constraint-induced movement therapy in stroke patients." *Neuroscience Letter,* 250(1):5–8.

Liu, D., J. Diorio, B. Tannenbaum, C. Caldji, D. Francis, A. Freedman, S. Sharma, D. Pearson, P. M. Plotsky, and M. J. Meaney. 1997. "Maternal Care, Hippocampal Glucocorticoid Receptors, and Hypothalamic-Pituitary-Adrenal Responses to Stress." *Science,* 277(5332):1659–62.

Loftus, E. 1993. "The reality of repressed memories." *American Psychologist,* 518–31.

Loftus, E., and K. Ketcham. 1994. *The Myth of Repressed Memory.* New York: St. Martin's Press.

Lorenz, K. 1973. "Autobiography." *Les Prix Nobel 1973.* The Nobel Foundation. Available at www.nobel.se/laureates/medicine-1973-2-autobio.html.

MacLean, P. D. 1973. *A Triune Concept of the Brain and Behavior.* Toronto: University of Toronto Press.

MacLean, P. D. 1985. "Brain evolution relating to family, play, and the separation call." *Archives of General Psychiatry,* 42: 405–17.

MacLean, P. D. 1990. *The Triune Brain in Evolution.* New York: Plenum Press.

Martin, L. J., D. M. Spicer, M. H. Lewis, J. P. Gluck, and L. C. Cork. 1991. "Social deprivation of infant rhesus monkeys alters the chemoarchitecture of the brain: I. Subcortical regions." *Journal of Neuroscience,* 11(11):3344–58.

Martone, M., N. Butters, M. Payne, J. T. Becker, and D. S. Sax. 1984. "Dissociations between skill learning and verbal recognition in amnesia and dementia." *Archives of Neurology,* 41: 965–70.

Maurois, A. 1963. *The New York Times,* April 14.

Mayburg, H. S., E. J. Lewis, W. Regenold, and H. N. Wagner. 1994. "Paralimbic hypoperfusion in unipolar depression." *Journal of Nuclear Medicine,* 35: 929–34.

McKenna, J. J. 1996. "Sudden Infant Death Syndrome in cross-cultural perspective: Is infant-parent cosleeping proactive?" *Annual Review of Anthropology,* 25: 201–16.

McKenna, J. J., S. Mosko, C. Dungy, and J. McAninch. 1990. "Sleep and arousal patterns of co-sleeping human mother/infant pairs: a preliminary physiological study with implications for the study of sudden infant death syndrome (SIDS)." *American Journal of Physical Anthropology,* 83(3):331–47.

McKenna, J. J., E. B. Thoman, T. F. Anders, A. Sadeh, V. L. Schechtman, and S. F. Glotzbach. 1993. "Infant-parent co-sleeping in an evolutionary perspective: implications for understanding infant sleep development and the sudden infant death syndrome." *Sleep,* 16(3):263–82.

McKinney, W. T. 1985. "Separation and depression: biological markers." In M. Reite and T. Field (eds.), *The Psychobiology of Attachment and Separation.* New York: Academic Press.

Meaney, M. J., D. H. Aitken, S. Bhatnagar, and R. M. Sapolsky. 1991. "Postnatal handling attenuates certain neuroendocrine, anatomical, and cognitive dysfunctions associated with aging in female rats." *Neurobiology of Aging,* 12: 31–8.

Mitchell, E. A., J. M. Thompson, A. W. Stewart, M. L. Webster, B. J. Taylor, I. B. Hassall, R. P. Ford, E. M. Allen, R. Scragg, and D. M. Becroft. 1992. "Postnatal depression and SIDS: a prospective study." *Journal of Paediatrics and Child Health,* 28(Supplement 1):S13–6.

Moon, C., R. P. Cooper, and W. P. Fifer. 1993. "Two-day-olds prefer their native language." *Infant Behavior and Development,* 16.

Murray, L. 1992. "The impact of postnatal depression on infant development." *Journal of Child Psychology and Psychiatry and Allied Disciplines,* 33(3):543–61.

Murray, L., P. J. Cooper, and A. Stein. 1991. "Postnatal depression and infant development." *British Medical Journal,* 302(6783):978–9.

Musset, Alfred de. 1965. "Souvenir." In M. Bishop (ed.), *A Survey of French Literature.* New York: Harcourt Brace Jovanovich.

Nabokov, V. 1955. *Lolita.* New York: G. P. Putnam's Sons.

Nabokov, V. 1980. *Lectures on Literature.* New York: Harcourt Brace Jovanovich.

Neisser, U., and N. Harsch. 1992. "Phantom Flashbulbs: False Recollections of Hearing the News About *Challenger.*" In E. Winograd and U. Neisser (eds.), *Affect and Accuracy in Recall: Studies of "Flashbulb" Memories.* New York: Cambridge University Press.

Nicholi, A. M. 1983. "The nontherapeutic use of psychoactive drugs. A modern epidemic." *New England Journal of Medicine*, 308(16):925–33.

Nissen, E., G. Lilja, A. M. Widstrom, and K. Uvnas-Moberg. 1995. "Elevation of oxytocin levels early post partum in women." *Acta Obstetricia et Gynecologica Scandinavica*, 74(7):530–3.

Nurco, D. N., T. W. Kinlock, K. E. O'Grady, and T. E. Hanlon. 1996. "Early family adversity as a precursor to narcotic addiction." *Drug and Alcohol Dependence*, 43(1–2): 103–13.

Nurco, D. N., T. W. Kinlock, K. E. O'Grady, and T. E. Hanlon. 1998. "Differential contributions of family and peer factors to the etiology of narcotic addiction." *Drug and Alcohol Dependence*, 51(3):229–37.

Ornish, D. 1998. *Love and Survival: The Scientific Basis for the Healing Power of Intimacy*. New York: HarperCollins.

Panksepp, J. 1994. "Evolution constructed the potential for subjective experience within the neurodynamics of the mammalian brain." In P. Ekman and R. J. Davidson (eds.), *The Nature of Emotion: Fundamental Questions*. New York: Oxford University Press.

Panksepp, J. 1992. "A critical role for 'affective neuroscience' in resolving what is basic about basic emotions." *Psychological Review*, 99(3):554–60.

Panksepp, J., S. M. Siviy, and L. A. Normansell. 1985. "Brain opioids and social emotions." In M. Reite and T. Field (eds.), *The Psychobiology of Attachment and Separation*. New York: Academic Press.

Park, D. 1988. *The How and the Why: An Essay on the Origins and Development of Physical Theory*. Princeton: Princeton University Press.

Pascal, B. *Pensées*. 1972. Paris: Librairie Générale Française.

Percy, W. 1992. "The Coming Crisis in Psychiatry." *Signposts in a Strange Land*. New York: Noonday Press.

Peredery, O., M. A. Persinger, C. Blomme, and G. Parker. 1992. "Absence of maternal behavior in rats with lithium-pilocarpine seizure-induced brain damage: support of MacLean's triune brain theory." *Physiology & Behavior*, 52: 665–71.

Perry, B. D., and D. Pollard. 1997. *Society For Neuroscience: Proceedings from Annual Meeting*, New Orleans.

Pinker, S. 1995. *The Language Instinct*. New York: HarperPerennial.

Pinker, S. 1997. *How the Mind Works*. New York: W. W. Norton and Company.

Pollak, S. 1999. University of Wisconsin, Madison, press release. Available at www.sciencedaily.com/releases/1999/04/990405065725.htm.

Raine, A., M. S. Buchsbaum, J. Stanley, S. Lottenberg, L. Abel, and J. Stoddard. 1994. "Selective reductions in prefrontal glucose metabolism in murderers." *Biological Psychiatry,* 36(6):365–73.

Rako, S., and H. Mazer (eds.). 1980. *Semrad: The Heart of a Therapist.* New York: Jason Aronson.

Reber, A. S. 1976. "Implicit learning of synthetic languages: the role of the instructional set." *Journal of Experimental Psychology: Human Learning and Memory,* 2: 88–94.

Redican, W. K. 1982. "An evolutionary perspective on human facial displays." In P. Ekman (ed.), *Emotion in the Human Face.* Second Edition. New York: Cambridge University Press.

Reite, M., and J. P. Capitanio. 1985. "On the nature of social separation and social attachment." In M. Reite and T. Field (eds.), *The Psychobiology of Attachment and Separation.* New York: Academic Press.

Remen, R. N., and D. Ornish. 1997. *Kitchen Table Wisdom: Stories That Heal.* New York: Riverhead Books.

Reynolds, I., and M. I. Rob. 1988. "The role of family difficulties in adolescent depression, drug-taking and other problem behaviours." *Medical Journal of Australia,* 149(5):250–6.

Ridley, M. 1993. "Neural Networking: Computer Prediction of Capital Markets." *The Economist,* October 9, p. 19.

Risser, D., A. Bönsch, and B. Schneider. 1996. "Family background of drug-related deaths: a descriptive study based on interviews with relatives of deceased drug users." *Journal of Forensic Science,* 41(6):960–2.

Robinson, D. 1992. "Implications of neural networks for how we think about brain function." *Behavioral and Brain Sciences,* 15: 644–55.

Rosenblum, L. A., and M. W. Andrews. 1994. "Influences of environmental demand on maternal behavior and infant development." *Acta Paediatrica Supplement,* 397: 57–63.

Rosenblum, L. A., J. D. Coplan, S. Friedman, T. Bassoff, J. M. Gorman, and M. W. Andrews. 1994. "Adverse early experiences affect noradrenergic and serotonergic functioning in adult primates." *Biological Psychiatry,* 35(4):221–7.

Ross, E. D. 1981. "The aprosodias: functional-anatomic organization of the affective components of language in the right hemisphere." *Archives of Neurology,* 38(9):561–9.

Rubinow, D. R., and R. M. Post. 1992. "Impaired recognition of affect in facial expression in depressed patients." *Biological Psychiatry,* 31(9):947–53.

Rumelhart, D. E., and J. L. McClelland. 1986. *Parallel Distributed Processing: Explorations in the Microstructure of Cognition.* Boston: MIT Press.

Russell, B. 1950. "An Outline of Intellectual Rubbish." *Unpopular Essays.* London: Routledge.

Russell, B. 1998. *Autobiography.* New York: Routledge.

Sakai, K., and Y. Miyashita. 1993. "Memory and imagery in the temporal lobe." *Current Opinion in Neurobiology,* 3(2):166–70.

Sander, L. W., G. Stechler, P. Gurns, and H. Julia. 1970. "Early mother-infant interaction and 24-hour patterns of activity and sleep." *Journal of the American Academy of Child Psychiatry,* 9: 103–23.

Schacter, D. L. 1990. "Memory." In M. Posner (ed.), *Foundations of Cognitive Science.* Boston: MIT Press.

Schiller, F. 1992. *Paul Broca: Explorer of the Brain.* London: Oxford University Press.

Selfridge, O. G. 1955. "Pattern recognition in modern computers." *Proceedings of the Western Joint Computer Conference.*

Selzer, R. 1982. *Letters to a Young Doctor.* New York: Simon & Schuster.

Shapiro, L. E., and T. R. Insel. 1990. "Infant's response to social separation reflects adult differences in affiliative behavior: a comparative developmental study in prairie and montane voles." *Developmental Psychobiology,* 23(5):375–93.

Shipley, W. V. 1963. "The demonstration in the domestic guinea-pig of a process resembling classic imprinting." *Animal Behavior,* 11: 470–4.

Singh, G. K., and S. M. Yu. 1996. "U.S. childhood mortality, 1950 through 1993: Trends and socioeconomic differentials." *American Journal of Public Health,* 86(4):505–12.

Smith, R. 1995. "The war on drugs: prohibition isn't working." *British Medical Journal,* 311(7021):1655–6.

Spiegel, D., J. R. Bloom, H. C. Kraemer, and E. Gottheil. 1989. "Effect of psychosocial treatment on survival of patients with metastatic breast cancer." *The Lancet,* 2(8668):888–91.

Spitz, R. 1945. "Hospitalism: an inquiry into the genesis of psychiatric conditions in early childhood." *Psychoanalytic Study of the Child,* 1: 53–74.

Spock, B. 1945. *Dr. Spock's Baby and Child Care.* New York: Meridith Press.

Squire, L. R., B. Knowlton, and G. Musen. 1993. "The structure and organization of memory." *Annual Review of Psychology*, 44: 453–95.

Stoker, A., and H. Swadi. 1990. "Perceived family relationships in drug abusing adolescents." *Drug and Alcohol Dependence*, 25(3):293–7.

Suomi, S. J. 1991. "Early stress and adult emotional reactivity in rhesus monkeys." *Ciba Foundation Symposium*, 156: 171–83.

Taylor, G. J. 1989. "Psychobiological Disregulation: A New Model of Disease." In *Psychosomatic Medicine and Contemporary Psychoanalysis*. Madison, Connecticut: International University Press.

Thoman, E. B., E. W. Ingersoll, and C. Acebo. 1991. "Premature infants seek rhythmic stimulation, and the experience facilitates neurobehavioral development." *Journal of Developmental and Behavioral Pediatrics*, 12(1):11–18.

Thomas, L. 1995. *Lives of a Cell: Notes of a Biology Watcher.* New York: Penguin.

Thoreau, H. D. 1960. "Where I Lived, and What I Lived For." In *Walden and Civil Disobedience.* New York: New American Library.

Tranel, D., and A. R. Damasio. 1993. "The covert learning of affective valence does not require structures in hippocampal system or amygdala." *The Journal of Cognitive Neuroscience*, 5(1):79–88.

Trevarthen, C. 1993. "The Self Born in Intersubjectivity: The Psychology of Infant Communicating." In U. Neisser (ed.), *The Perceived Self: Ecological and Interpersonal Sources of Self-knowledge.* New York: Cambridge University Press.

Truzzi, M. 1983. "Sherlock Holmes: Applied Social Psychologist." In U. Eco and T. A. Sebeok (eds.), *The Sign of Three: Dupin, Holmes, Pierce.* Bloomington: Indiana University Press.

Tucker, D. M., and P. Luu. 1998. "Cathexis revisited: corticolimbic resonance and the adaptive control of memory." *Annals of the New York Academy of Sciences*, 843: 134–52.

Viinamki, H., J. Kuikka, J. Tiihonen, and J. Lehtonen. 1998. "Change in monoamine transporter density related to clinical recovery: a case-control study." *Nordic Journal of Psychiatry*, 52: 39–44.

Wallerstein, J. S., and J. B. Kelly. 1980. *Surviving the Breakup: How Children and Parents Cope with Divorce.* New York: Basic Books.

Ward, A. 1948. "The cingular gyrus: Area 24." *Journal of Neurophysiology*, 11: 13–23.

Watson, J. B. 1928. *Psychological Care of the Infant and Child.* New York: W. W. Norton.

Wilson, E. O. 1998. *Consilience: The Unity of Knowledge.* New York: Knopf.

Wise, S. P., and M. Herkenham. 1982. "Opiate receptor distribution in the cerebral cortex of the rhesus monkey." *Science,* 218: 387–9.

Wolfe, G. 1994. *Shadow and Claw.* New York: Tom Doherty Associates.

Wright, R. 1997. "The urge to let a child fall asleep in your bed is natural. Surrender to it." *Slate,* March 27, 1997.

Yeats, W. B. 1996. "Among Schoolchildren." In R. J. Finneran (ed.), *The Collected Poems of W. B. Yeats.* New York: Simon & Schuster.

Zeskind, P. S., S. Parker-Price, and R. G. Barr. 1993. "Rhythmic organization of the sound of infant crying." *Developmental Psychobiology,* 26(6):321–33.

Zwillich, T. 1999. "Interpersonal psychotherapy can normalize brain: neuroimaging shows therapy can return mood regulation structures to normal." *Clinical Psychiatry News,* 27(7):1, 6.

ACKNOWLEDGMENTS

Every book comes to life in that luminous place wherein minds intersect and hearts meet. From the first germ of an idea to the neat flutter of pages a reader holds in his hand, a book owes its existence to the collaborative efforts of a dedicated band—and this work, with its trio of primary collaborators, more so than most.

In 1991, when I encountered Dr. Amini and Dr. Lannon at the University of California, San Francisco, the former had been there for twenty-eight years and the latter for twelve; they had been working together since 1970. Between them they had treated thousands of patients and taught hundreds of physicians and therapists-in-training. I was in my training residency at the time, reeling under the bombardment of contradictory and dubious doctrines that constitutes a modern psychiatric education. While I knew little about psychiatry and less about people, I was able to recognize that the seminars Dr. Amini and Dr. Lannon led were remarkable—a psychoanalyst and a biological psychiatrist teaming up, with amity and respect, to discuss psychotherapy, development, mood disorders, and love. Readers unfamiliar with the divisive quality of academic psychiatry may not appreciate the rarity of such a pairing—comparable, say, to finding a Montague and a Capulet taking turns quaffing beer out of a single stein in the local pub.

That notable conjunction was only the beginning of the unexpected. The richly complex seminars that Drs. Amini and Lannon wove were not easy for the intellect to grasp or words to frame. A good many residents were dazed and baffled. At first blush, I took my teachers' apparent and maddening imprecision to be evidence of a certain vagueness of thought fairly epidemic in the field. Eventually, with a dawning sense of wonder, I realized that my incomprehension represented not ineptitude on anyone's part, but necessity—the two were speaking from another plane, as it were, about matters not readily translated into words. I came to understand that in the course of their clinical careers (long before mine began), Drs. Amini and Lannon had accumulated an impressive store of wisdom about the ways of the human heart. They steadfastly resisted the fatal simplifications and seductively pat solutions that we residents demanded of them daily, because they were certain that such satisfying shortcuts would be meaningless. Instead, they actually tried to impart to their students the hard-won secrets that are expressible, as music is, in a language very different from one to which people are most accustomed.

This approach to psychiatry and to the teaching of psychiatry was wholly novel; to this day I have encountered nothing like it anywhere. I decided then that if I was to learn what a psychiatrist—or anybody—needs to know about the human heart, I would learn it from Drs. Amini and Lannon or not at all.

After I graduated from the residency, Dr. Amini and Dr. Lannon permitted me to join them in teaching the same seminars in which I had first encountered that resonant dimension wherein emotional life takes place. A successful two-party collaboration is difficult enough to manage, and one among three people, more precarious still. Nevertheless, our triumvirate proved itself the improbable miracle. When we combined our ideas and energies, ideas flew fast and thick. Most revolved around the centerpiece of Dr. Amini's clinical work: a conviction of the life-shaping force inherent in emotional contact between two minds.

We set about learning all that we could about the biological reality of that elusive and powerful phenomenon. Pursuing such a project required more knowledge than any of us possessed. And so for years, we collected every fact relevant to emotional life, and we met on Saturday mornings to pore over the findings. Over orange juice and muffins and eggs, we taught one another arcane information from unfamiliar schools, and we listened, explored, and argued. The stray elements slowly began to cohere. Their ultimate synthesis finds its expression in this book.

Much has transpired between those weekend colloquys and the book you hold in your hands. Early incarnations of our ideas were tested on successive waves of UCSF psychiatric residents. On the whetstone of their affable skepticism, we honed our concepts. We delivered presentations to professional audiences at UCSF Grand Rounds, the San Francisco Psychoanalytic Institute, and at the annual meetings of the American Group Psychotherapy Association and the American Psychiatric Association (joined, in most such outings, by friend and colleague Dr. Alan Louie). The resulting professional dialogues assisted us in further defining our thoughts and our mission.

When we had achieved a certain minimal level of coherence, we published a compact summary of our paradigm in the journal *Psychiatry*. The masochistic reader who unearths the relevant issue of that benevolent journal may find a few glimmers of our ideas visible there, peeking out from under the crushing density of academic language. When we departed the university in 1996—first Dr. Amini, then Dr. Lannon, and finally myself—we resolved to devote the time thus freed to setting down our theory of love, in readable form, within the pages of a book.

When I set out to distill our shared ideas into words, to give them a voice and a form, I encountered a formidable obstacle: I did not know how to write well enough to succeed in that endeavor. "No invective can adequately excoriate this dreadful book," I read some decades ago, in a review of a now forgotten work. It is my sad duty to report that this judgment applied equally well to the first draft of the manuscript that I produced in attempting to translate these ideas into a verbal language. The book would have foundered then and there, on the shoals of my ignorance, had I not been fortunate enough to encounter Dr. Glenda Hobbs. I am deeply grateful to her for teaching me nearly all that I know about writing. Ages ago, the Chinese philosopher Lao-tzu wrote, "To see things in the seed—that is genius." In this instance that credit belongs to Dr. Hobbs—who, by means of an alchemical art I cannot fathom, discerned a valuable core behind the dross of verbiage that constituted this book's first draft and was instrumental in bringing the best to the fore.

A number of people were kind enough to read permutations and portions of the manuscript, in its various stages of undress, and offer their useful comments: Liz Amini, Cecilia Lannon, Sue Lewis, Christina Amini, Alan Langerman, Linda Motzkin, Sue Halpern, Caryl Gorska, David Berlinski, Robert DiNicolantonio, Aimee West, Ezra Epstein, Kristen Engle, Maia Wilkinson, Mike Meixel, and Deb Seymour. My good friend and colleague Ed Burke was particularly generous in his willingness to read countless versions without complaint, and he lent me not only his keen eye but also his voluminous knowledge on every conceivable subject. Mark Powelson's encouragement and guidance took a novice through the vast, thicketed wilderness of inexperience and up to the very gates of the publishing industry. Our heartfelt thanks to Lisa Motzkin of Motzkin Design, whose lovely art graces this book's interior. Paul Ekman, Robert Wallerstein, and Judith Wallerstein, longtime friends and colleagues, have been valiant supporters of this book, and we appreciate their indispensable efforts on its behalf.

Drs. Wallerstein and Wallerstein were kind enough to speed us on to our agent, Carol Mann, to whom we are profoundly indebted. Her advice and guidance on innumerable matters have been invaluable. While we had hoped that we could interest an agent in our work, we never dreamed that we would encounter someone whose grasp of and enthusiasm for the book exceeded, in some respects, our own.

The same is true of the devoted team at Random House, where so many people have treated this book with loving care; to all of them, our deepest thanks. Scott Moyers's ardent dedication to this project has been a crucial inspiration. He wielded his editorial blade with a surgeon's precision and an artist's finesse, and spurred me on to reach for the better book he believed in, lurking beneath the surface of the one I had shown him. We could not have found more able or careful hands into which to entrust the fruit of so many years' labor. Kate Niedzwiecki provided tireless assistance on matters too diverse and numerous to tabulate. Benjamin Dreyer, assisted by Jennifer Prior, eradicated from the text a multitude of missteps, major and minor, and guided the book through production with consummate skill and concern. The loveliness of Andy Carpenter's jacket design speaks volumes for itself, in its own silent and evocative medium. The passionate support of Random's deputy publisher, Mary Bahr, has been exhilarating and pivotal. Our sincere thanks to Ann Godoff, Random House's president, publisher, and editor in chief, for her unwavering enthusiasm and vital creative input.

Our patients have taught us much, and we owe them a debt of gratitude for the privilege of inviting us into their lives and for their courage to meet us in the arduous challenge of change. Finally, we wish to thank our respective families, without whose love and forbearance none of this could have proved possible.

In assembling this account of the causative factors, the formative agents that impinged on the book from prehistory to final configuration, I am struck by the improbability of the whole enterprise—that *these* words should be assembled in *this* order within these covers. Without the participation and fervor of each person enumerated above (and many others who are not so listed) there would have been no book, or, at least, a very different and inferior one. On just such a fragile chain of coincidence does every life hang—leaving all the more reason for joy and celebration when, against apparently insurmountable odds, matters turn out right.

Thomas Lewis, M.D.
Sausalito, California

INDEX

Page numbers in *italics* refer to illustrations and tables.